SpringerBriefs in Electrical and Computer Engineering

Series editors
Woon-Seng Gan
Sch of Electrical & Electronic Engg
Nanyang Technological University
Singapore, Singapore

C.-C. Jay Kuo
University of Southern California
Los Angeles, California, USA

Thomas Fang Zheng
Res Inst Info Tech
Tsinghua University
Beijing, China

Mauro Barni
Dept of Info Engg & Mathematics
University of Siena
Siena, Italy

Xiaofan He • Huaiyu Dai

Dynamic Games for Network Security

 Springer

Xiaofan He
Department of Electrical Engineering
Lamar University
Beaumont, TX, USA

Huaiyu Dai
Department of Electrical and Computer
Engineering
North Carolina State University
Raleigh, NC, USA

ISSN 2191-8112   ISSN 2191-8120   (electronic)
SpringerBriefs in Electrical and Computer Engineering
ISBN 978-3-319-75870-1   ISBN 978-3-319-75871-8   (eBook)
https://doi.org/10.1007/978-3-319-75871-8

Library of Congress Control Number: 2018933373

Printed on acid-free paper

This Springer imprint is published by the registered company Springer International Publishing AG part of Springer Nature.
The registered company address is: Gewerbestrasse 11, 6330 Cham, Switzerland

To my beloved family.

Xiaofan He

To my parents and my family.

Huaiyu Dai

Preface

The recent emergence and advancement of various information and cyber-physical networks have brought unprecedented convenience to our daily lives. To ensure effective and continuous operations of these modern networks, it is of crucial importance to deploy efficient and reliable defense mechanisms to protect their security. However, in the security battles, one challenge is that the adversary is constantly upgrading their attacking tactics and becoming increasingly intelligent, making conventional static security mechanisms outdated and incompetent. Considering this, game theory, which is a rich set of analytic tools for modeling and analyzing the strategic interactions among intelligent entities, has been widely employed by the network security community for predicting the adversary's attacking strategy and designing the corresponding optimal defense. Despite its celebrated applications in addressing some network security problems, the classic game theory mainly focuses on static settings, while many practical security competitions often take place in dynamic scenarios due to frequent changes in both the ambient environment and the underlying networks. This motivates the recent exploration of the more advanced stochastic game (SG) theory that can capture not only the interactions between the defender and the attacker but also the environmental dynamics. The objective of this book is to collect and systematically present the state of the art in this research field and the underlying game-theoretic and learning tools to the broader audience with general network security and engineering backgrounds.

Our exposition of this book begins with a brief introduction of relevant background knowledge in Chap. 1. Elementary game theory, Markov decision process (MDP), and SG are covered, including the basic concepts and mathematical models as well as the corresponding solution techniques. With this necessary background, in Chap. 2, we proceed to review existing applications of SG in addressing various dynamic security games, in the context of cyber networks, wireless networks, and cyber-physical networks. In these applications, the defenders and the attackers are assumed to hold equal information about the corresponding security competitions, whereas information asymmetry often exists in practice. Considering this, we take a step further and explore how to deal with dynamic security games in the presence of information asymmetry in Chaps. 3–5. In particular, our exploration includes three aspects of this issue—dynamic security games with extra information, dynamic security games with incomplete information, and dynamic security games with deception. It is worth mentioning that, although we mainly take the defender's perspective in the discussions, the corresponding results and techniques may be employed to predict the attacker's behavior in similar situations. More specifically, dynamic security games with extra information discussed in Chap. 3 concern security competitions where the defender has an informational advantage over the adversary. Based on the existing SG framework, we present a novel technique that enables the defender to fully exploit such advantage so as to achieve faster adaptation and learning in dynamic security competitions. The complementary scenarios where the defender lacks information about the adversary are examined in Chap. 4 through the lens of incomplete information SG. To address incomplete information SGs, a new algorithm that integrates Bayesian learning and conventional learning algorithms of SG is presented; the key idea is to allow the defender to gradually infer the missing information through repeated interactions with the adversary. The extra and the incomplete information considered in Chaps. 3 and 4 is inherent to the corresponding security problems. In Chap. 5, we switch gear and further explore how to proactively create information asymmetry for the defender's benefit, and the dynamic deception technique is investigated as an effective tool to achieve this objective. Lastly, concluding remarks and our perspective for future works are presented in Chap. 6.

The authors would like to acknowledge Prof. Rudra Dutta, Prof. Peng Ning, and Mr. Richeng Jin. Without their contribution, this book could not have been made possible. We would also like to thank all the colleagues and researchers for their pioneering and inspiring works that lay out the solid foundation of this book.

Wuhan, Hubei, China   Xiaofan He
Raleigh, NC, USA   Huaiyu Dai

Contents

Chapter 1
Preliminaries

1.1 Introduction

To start our journey in this book, relevant backgrounds on game theory, Markov decision process (MDP), and stochastic game (SG) will be introduced first in this chapter to pave the way for our later exposition.

In essence, many security issues can be treated as a game between the defender and the attacker who are intelligent entities that can smartly plan their actions in the security rivalries. For this reason, our discussion in this chapter begins with the presentation of some rudimentary concepts of the classic game theory—a fascinating theory that concerns the modeling and prediction of the behaviors of intelligent entities in strategical interactions. In particular, the basic elements of a game will be introduced first. Then, the Nash equilibrium (NE), one of the most widely adopted solution concept in classic game theory, will be reviewed. Besides, several important analytic results regarding the existence and uniqueness of the NE are also discussed.

In addition to the interactions with the opponent, practical security problems often involve with different environmental, system, or network dynamics. As a fundamental tool for optimal planning in dynamic environments, the MDP is also reviewed in this chapter. Specifically, two key notions for the MDP, the Q-function and the value function, are introduced first, along with the Bellman's principle of optimality. To solve the MDP, several computing methods that can be used to directly derive the optimal solution when full information of the MDP is available are presented. To deal with situations with unknown information, the Q-learning algorithm that enables the agent to gradually adjust its strategy based on repeated interactions with the dynamic environment is illustrated.

As a key framework for addressing various dynamic security games, the SG will be reviewed in this chapter as well. The SG is a natural marriage of the classic

X. He, H. Dai, *Dynamic Games for Network Security*, SpringerBriefs in Electrical and Computer Engineering, https://doi.org/10.1007/978-3-319-75871-8_1

game theory and the MDP and therefore can jointly manage the interactions with the opponent and the environmental dynamics, well suited to practical security problems. With this consideration, the underlying model of SG will be presented in details, and whenever applicable, its connections to the classic game theory and the MDP will be drawn. Similar to the case of MDP, both the computing methods and the learning algorithms for SG are presented, including the non-linear programming method, value iteration method, the minimax-Q algorithm, the Nash-Q algorithm and the Win-or-Learn-Fast (WoLF) algorithm.

1.2 Elementary of Game Theory

Game theory [1, 2] is the study of the strategic interactions among multiple intelligent entities, termed *players* hereafter. The objective is to acquire certain reasonable prediction about the behaviors of the entities and help them design adequate strategies that can lead to the best possible benefits. In some sense, game theory can be treated as competitive optimization problems in which the players have to consider the influence from the others when optimizing their own payoffs. Although originated as a celebrated branch of mathematics and economics, game theory also finds its success in addressing various engineering problems, such as resource allocation for wireless networks [3], patrolling strategy design for airport security [4], defense system configuration in cyber-networks [5], among the many others. It is such a profound subject that covers a large variety of different topics, and each of them deserves a separate book. Due to space limitation, in this book, we only provide a very brief and rudimentary introduction to this fascinating theory so as to prepare readers lacking relevant backgrounds for our later expositions. For more comprehensive and systematic treatments of classic game theory, interested readers may refer to [1, 2] and other relevant materials.

Roughly speaking, a game is a procedure of interactive decision-making among multiple players. Although some slight differences may exist, several basic elements are shared by almost all types of games, and will be introduced in the sequel. As already mentioned above, the set $\mathscr{I}$ of players is one of the elementary building block of a game $\mathscr{G}$. In the game, each player-i ($i \in \mathscr{I}$) takes an *action* a^i from its action set $\mathscr{A}^i$. Based on the actions taken by the players, each player-i will receive a *payoff* $r^i = R^i(a^i, a^{-i})$ (or sometimes interchangeably called *reward*), with $R^i(\cdot)$ the *payoff function*. Here, we follow the convention of using a^{-i} to represent the actions from all the players other than player-i, i.e., $a^{-i} \triangleq \left(a^1, \ldots, a^{i-1}, a^{i+1}, \ldots, a^{|\mathscr{I}|}\right)$. The objective of each player is to maximize its own payoff r^i by properly selecting an action a^i. The underlying policy according to which a player takes its action is called the *strategy* of that player, and is often denoted by π^i. Two types of strategies are often considered in game theory: *pure strategy* and *mixed strategy*. A pure strategy directly specifies which action to take. For example, in a rock-paper-scissors game, "play rock" is a pure strategy that informs the player to always take the action "rock". A mixed strategy is often represented by a probability distribution over the action set and only specifies in a probabilistic manner about which action to take. Taking the rock-paper-scissors game as an example again, the mixed strategy $\pi^i = [\frac{1}{2}, \frac{1}{4}, \frac{1}{4}]$ dictates that player-i will play rock with probability $\frac{1}{2}$ and play paper and scissors with equal probability $\frac{1}{4}$. In general, a mixed strategy of player-i can be written as a $|\mathscr{A}^i|$-dimensional vector $\pi^i = [p^i_1, \ldots, p^i_{|\mathscr{A}^i|}]$ in which p^i_j (for $1 \leq j \leq |\mathscr{A}^i|$) represents the probability of taking the jth action from the set $\mathscr{A}^i$; clearly, $\sum\limits_{j=1}^{|\mathscr{A}^i|} p^i_j = 1$. In addition, it is worth mentioning that any pure strategy can be expressed as a mixed strategy. For example, the pure strategy "play rock" considered above can be written as $\pi^i = [1, 0, 0]$. Nonetheless, as the players often do not have prior information about which actions will be taken by the others, it is highly non-trivial for each player to find the "optimal" strategy. In fact, unlike conventional optimization problems concerned with optimality, game theory considers a different solution concept—*equilibrium*; this is mainly because the players in a game are assumed to be *self-interested*, caring only about their own payoffs. Although many different notions of equilibrium have been developed in literature [1, 2], Nash equilibrium (NE) [6, 7] is probably the most widely adopted one, and many fundamental results in game theory are centered around this concept. Specifically, the NE is defined as follows.

Definition 1 A tuple of strategies $\left(\pi^1_*, \ldots, \pi^{|\mathscr{I}|}_*\right)$ form an *NE* if the following condition holds: For any player-i and strategy π^i, it always has

$$\mathbb{E}_{\left(\pi^1_*, \ldots, \pi^i_*, \ldots \pi^{|\mathscr{I}|}_*\right)} \left[R^i(a^i, a^{-i})\right] \geq \mathbb{E}_{\left(\pi^1_*, \ldots, \pi^i, \ldots \pi^{|\mathscr{I}|}_*\right)} \left[R^i(a^i, a^{-i})\right], \quad (1.1)$$

where on the left-hand side, the subscript $\left(\pi^1_*, \ldots, \pi^i_*, \ldots \pi^{|\mathscr{I}|}_*\right)$ indicates that the expectation is taken according to the law determined by this strategy tuple, and similar notation is used on the right-hand side.

Intuitively, the above definition says, a strategy tuple is an NE if no player can increase its expected payoff by *unilaterally* changing its strategy.

Several fundamental results regarding the existence and uniqueness of NE are presented below with the corresponding proofs omitted.

Theorem 1.1 ([6]) *Every finite game in which the numbers of players and actions are finite (i.e., $|\mathscr{I}| < \infty$ and $|\mathscr{A}^1|, \ldots, |\mathscr{A}^{|\mathscr{I}|}| < \infty$) has a mixed strategy NE.*

Theorem 1.2 ([8]) *For an infinite game with $|\mathscr{I}| < \infty$, if*

(i) the action spaces $\mathscr{A}^i$'s are nonempty and compact metric spaces;
(ii) the payoff functions $R^i(a^i, a^{-i})$'s are continuous;

then there always exists a mixed strategy NE. If

  (i) *the action spaces $\mathscr{A}^i$ are compact and convex;*
  (ii) *the payoff functions $R^i(a^i, a^{-i})$'s are continuous in a^{-i};*
  (iii) *the payoff functions $R^i(a^i, a^{-i})$'s are continuous and (quasi-)concave in a^{-i};*

then there always exists a pure strategy NE.

Theorem 1.3 ([9]) *For an infinite game with $|\mathscr{I}| < \infty$, if*

  (i) *the pure strategy set of each player-i is specified in the form*

$$\Pi^i = \left\{ \pi^i \,|\, f_i(\pi^i) \geq 0 \right\}, \tag{1.2}$$

  for some concave function $f_i()$, and for each $i \in \mathscr{I}$, there exists at least a point x_i such that $f_i(x_i)$ is strictly positive;
  (ii) *the payoff functions $(R^1, \ldots, R^{|\mathscr{I}|})$ are diagonally strictly concave over the set $\Pi^1 \times \cdots \times \Pi^{|\mathscr{I}|}$, in which Π^i is the set of all possible strategies of player-i;*

then the game has a unique pure strategy NE.

1.3 The Markov Decision Process

1.3.1 The MDP Model

Before introducing the SG, the single-player version of SG, known as the MDP [10], is reviewed in this section to better prepare the readers for our later expositions. As depicted in Fig. 1.1, an MDP concerns the interactions between a dynamic system or environment and an intelligent agent. Particularly, an MDP unfolds as follows. At the beginning of each timeslot n, the agent first observes the current state $s_n \in \mathscr{S}$ of the system (with $\mathscr{S}$ the set of possible states) and then takes an a_n chosen from its action set $\mathscr{A}$ according to its strategy π. The strategy of the agent maps a state $s \in \mathscr{S}$ into a probability distribution over the action set $\mathscr{A}$. More specifically, $\pi(s, a)$ will be used to denote the probability of the agent taking action a in state s; clearly $\sum_{a \in \mathscr{A}} \pi(s, a) = 1$. After this, on the one hand, the agent will receive a reward $r_n = R(s_n, a_n)$, where $R()$ is the state dependent payoff function associated with this MDP. On the other hand, the system will transit into a new state s_{n+1} dictated by a controlled Markov process with transition probability $\mathbb{P}(s_{n+1}|s_n, a_n)$. Then, this process repeats. The objective of the agent in the MDP is to maximize the expected accumulative long-term reward, which is often expressed as $\mathbb{E}\{\sum_{n=0}^{\infty} \beta^n \cdot r_n\}$. Here, $0 \leq \beta < 1$ is the discounting factor of the MDP. More specifically, $\beta = 0$ corresponds to a myopic agent that only cares about its instant reward, while a non-zero β indicates that the agent concerns its long-term performance but puts decreasing emphases for the rewards obtained in the further future due to the increasing uncertainty. In the rest of this book, we will mainly focus on scenarios with $\beta > 0$. In such scenarios, since the agent's action influences not only its instant reward r_n but also the state transition and thus the future rewards, the agent has to conduct *foresighted optimization* when designing its strategy π.

Fig. 1.1 Diagram of MDP

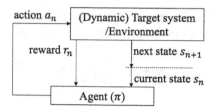

To find the agent's optimal strategy in an MDP, the *Q-function* and the *value function* are defined as follows

$$Q_*(s, a) \triangleq \mathbb{E}\left[R(s, a) + \beta \cdot V_*(s')\right]$$

$$= R(s, a) + \beta \cdot \sum_{s' \in \mathscr{S}} \mathbb{P}(s'|s, a)V_*(s'), \qquad \forall\,(s, a) \in \mathscr{S} \times \mathscr{A}, \quad (1.3)$$

and

$$V_*(s) \triangleq \max_{a' \in \mathscr{A}} Q_*(s, a'), \qquad \forall\, s \in \mathscr{S}. \tag{1.4}$$

Some interpretations of these functions are in order. The optimal value function $V_*(s)$ represents the best possible expected accumulative long-term reward that can be achieved by the agent given an initial state s. The optimal Q-function $Q_*(s, a)$ corresponds to the long-term reward obtained when the agent takes action a at the given initial state s and then follows the optimal strategy afterwards. The above result is known as the *Bellman equation* [11]. Given the optimal Q-function, the agent can readily derive the optimal strategy $\pi_*(s)$ for each state s as follows

$$\pi_*(s) = \arg\max_{\pi} \sum_{a} Q_*(s, a)\pi(s, a) = \arg\max_{a} Q_*(s, a), \tag{1.5}$$

where the last equality holds as it can be readily shown that the optimal strategy in MDP is always a pure strategy.

1.3.2   Solving the MDP

With the above discussion, one may realize that solving an MDP can be eventually boiled down to finding the optimal Q-function. Depending on the information available to the agent, there are several ways to compute the Q-function as introduced below, but by no means this is a complete list. Particularly, when the state transition function $\mathbb{P}(s'|s, a)$ and the reward function $R(s, a)$ are known to the agent, the agent can adopt the linear programming method or the value iteration method [12]. When such information is unknown, the Q-learning algorithm [13] can be adopted.

Linear Programming In this approach, the agent first solves the following linear programming problem to find the optimal value functions of the MDP.

$$\min_{\{V(s)\}_{s \in \mathscr{S}}} \frac{1}{|\mathscr{S}|} \sum_{s \in \mathscr{S}} V(s) \qquad \text{(LP)}$$

$$\text{s.t.} \quad V(s) \geq R(s, a) + \beta \cdot \sum_{s'} \mathbb{P}(s'|s, a)V(s'), \ \forall s \in \mathscr{S}, a \in \mathscr{A}.$$

Then, the optimal Q-function can be found through (1.3).

Value Iteration In this approach, the agent starts with an arbitrary initial values and executes the following two iterative steps until convergence.

$$Q_{n+1}(s, a) = R(s, a) + \beta \cdot \sum_{s'} \mathbb{P}(s'|s, a)V_n(s'), \ \text{for all } s, a, \qquad (1.6)$$

and

$$V_{n+1}(s) \triangleq \max_{a'} Q_n(s, a'), \ \text{for all } s. \qquad (1.7)$$

Note that the convergence of the value iteration approach is guaranteed when β is strictly less than one. Variants of value iteration are also available in literature, including the policy iteration method [10] and the modified policy iteration [14].

Q-Learning This approach allows the agent to gradually learn the Q- and the value functions through repeated interactions with the dynamic system, without requiring any prior information about the state transition probability and the reward function. Specifically, at each timeslot n, the agent observes the current state s_n and then takes an action a_n specified by its current strategy π_n. Then, based on the received reward r_n and the observed state transition from s_n to s_{n+1}, the agent updates its Q- and value functions as follows

$$Q_{n+1}(s, a) = \begin{cases} (1 - \alpha_n)Q_n(s, a) + \alpha_n[r(s, a) + \beta V_n(s_{n+1})], \\ \qquad\qquad\qquad\qquad \text{if } (s, a) = (s_n, a_n), \qquad (1.8) \\ Q_n(s, a), \qquad\qquad\qquad\qquad \text{otherwise}, \end{cases}$$

$$V_{n+1}(s) = \max_a Q_{n+1}(s, a), \qquad (1.9)$$

where α_n is the learning rate that admits the following standard conditions of reinforcement learning [15]

$$0 \le \alpha_n < 1, \quad \sum_{n=0}^{\infty} \alpha_n = \infty, \quad \sum_{n=0}^{\infty} \alpha_n^2 < \infty. \tag{1.10}$$

Then, based on the updated Q-function, the agent updates its strategy as follows

$$\pi_{n+1}(s) = \arg\max_a Q_{n+1}(s, a). \tag{1.11}$$

The convergence of the Q-learning algorithm has been established in [13].

1.4   The Stochastic Games

1.4.1   The Model of SG

The SG is an extension of the MDP to the more general multi-agent settings. As the focus of this book is on security games, the following discussions will be mainly devoted to two-player SGs, in which one player corresponds the defender and the other corresponds to the attacker. Such model is suitable to multi-defender and multi-attacker scenarios as well when a central controller exists on both sides. In addition, one may find that the models and techniques presented below can be readily extended to the cases of distributed defenders and attackers.

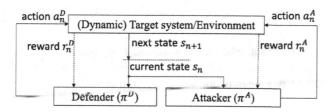

Fig. 1.2 Diagram of SG

As depicted in Fig. 1.2, a typical two-player SG [16] consists of two players and a dynamic system (or environment). The interactions among these three components unfold as follows. At the beginning of timeslot n, both players observe the current state s_n of the dynamic system. Based on the observed state and their own strategies π^D and π^A, they take certain actions a_n^D and a_n^A, respectively. Depending on the actions and current system state, both players will receive their rewards, denoted by r_n^D and r_n^A, determined by their reward functions $R^D(s, a^D, a^A)$ and $R^A(s, a^D, a^A)$, respectively. In the meantime, the system evolves into a new state s_{n+1} with a controlled probability $\mathbb{P}(s_{n+1}|s_n, a_n^D, a_n^A)$. Similar to the MDP, the objectives of the players are to optimize their average accumulative long-term rewards given by $\mathbb{E}\left[\sum_{n=1}^{\infty} \beta^n \cdot r_n^D\right]$ and $\mathbb{E}\left[\sum_{n=1}^{\infty} \beta^n \cdot r_n^A\right]$ (for $\beta \in [0, 1)$), respectively.[1]

A fundamental difference between SG and MDP is that, in SG, the players have to consider not only a foresighted optimization problem as in MDP but also the potential influence from the other player. Therefore, one has to combine relevant concepts and techniques from MDP and game theory to devise a solution to the SG. To this end, we first extend the definitions of the Q- and the value functions in MDP to the SG settings. Particularly, the optimal Q-function of player-I (or, equivalently, the defender) is defined as

$$Q_*^D(s, a^D, a^A) \triangleq \mathbb{E}[R^D(s, a^D, a^A) + \beta V_*^D(s')], \tag{1.12}$$

and the Q-function $Q_*^A(s, a^D, a^A)$ of player-II (or, equivalently, the attacker) is similarly defined. Unlike in MDP, some extra efforts are required to define the value function in SG. Specifically, let $\text{NASH}^D(X, Y)$ be the operator of finding the value of the first player in a two-player game at an NE, where the payoff matrices of the first and the second players are denoted by X and Y, respectively. In addition, let $\arg \text{NASH}^D(X, Y)$ be the operator of finding the strategy of the first player at an NE. With these notations, one may define the optimal value function of the defender as

$$V_*^D(s) \triangleq \text{NASH}^D\left(Q_*^D(s, \cdot, \cdot), Q_*^A(s, \cdot, \cdot)\right), \tag{1.13}$$

and the corresponding NE strategy at state s is given by

$$\pi_*^D(s) \triangleq \arg \text{NASH}^D\left(Q_*^D(s, \cdot, \cdot), Q_*^A(s, \cdot, \cdot)\right). \tag{1.14}$$

It is worth mentioning that, when the reward functions admit a zero-sum structure, i.e., $R^D(s, a^D, a^A) = -R^A(s, a^D, a^A)$, the $\text{NASH}^D(\cdot, \cdot)$ operator reduces to the max min operator. Consequently, the value function admits

$$V_*^D(s) = \max_{\pi(s,\cdot)} \min_{a^A} \sum_{a^D} Q_*^D(s, a^D, a^A)\pi(s, a^D), \tag{1.15}$$

and the corresponding strategy is given by

$$\pi_*^D(s) \triangleq \arg \max_{\pi(s,\cdot)} \min_{a^A} \sum_{a^D} Q_*^D(s, a^D, a^A)\pi(s, a^D). \tag{1.16}$$

---

[1] In this book, we mainly focus on the discounted SG where $\beta < 1$ and the SGs concern the limiting average reward are beyond the scope of this book.

1.4.2 Solving the SG

As the case of MDP, depending on the information available to the players, different methods have been developed in literature to solve the SG. In the sequel, several representative ones will be introduced, and interested readers may refer to, for example, [16–18] for more comprehensive surveys.

Specifically, when the state transition probabilities and the reward functions are known to the players, the non-linear programming method and the value iteration method can be employed to solve the SG [12].

Non-linear Programming Before demonstrating the nonlinear programming method, several new notations are introduced first to smooth the discussion. First, define the defender's instant reward matrix

$$\mathbf{R}^D(s) = \left[R^D(s, a^D, a^A) \right]_{a^D \in \mathscr{A}^D, a^A \in \mathscr{A}^A}, \tag{1.17}$$

and then, for a given value function V^D, define a future reward matrix $\mathbf{T}$

$$\mathbf{T}(s, V^D) = \left[\sum_{s' \in \mathscr{S}} \mathbb{P}(s'|s, a^D, a^A) V^D(s') \right]_{a^D \in \mathscr{A}^D, a^A \in \mathscr{A}^A}, \tag{1.18}$$

Additionally, for a given strategy pair π^D and π^A, define a vector

$$\mathbf{r}^D(\pi^D, \pi^A) = \left[\left(\pi^D(s, \cdot) \right)^T \mathbf{R}^D(s) \pi^A(s, \cdot) \right]_{s \in \mathscr{S}}, \tag{1.19}$$

and a state transition matrix

$$\mathbf{P}(\pi^D, \pi^A) = \left[\mathbb{P}(s'|s, \pi^D, \pi^A) \right]_{s, s' \in \mathscr{S}}. \tag{1.20}$$

With these quantities, the following nonlinear programming can be formulated to find an NE of the SG:

$$\min_{V^D, V^A, \pi^D, \pi^A} \quad \left(V^D - \mathbf{r}^D(\pi^D, \pi^A) - \beta \mathbf{P}(\pi^D, \pi^A) V^D \right) \tag{NLP}$$

$$+ \left(V^A - \mathbf{r}^A(\pi^D, \pi^A) - \beta \mathbf{P}(\pi^D, \pi^A) V^A \right)$$

$$\text{s.t.} \quad \mathbf{R}^D(s) \pi^A(s, \cdot) + \beta \cdot \mathbf{T}(s, V^D) \pi^A(s, \cdot) \leq V^D(s) \mathbf{1}, \ \forall s \in \mathscr{S}$$

$$\pi^D(s, \cdot) \mathbf{R}^A(s) + \beta \cdot \pi^D(s, \cdot) \mathbf{T}(s, V^A) \leq V^A(s) \mathbf{1}, \ \forall s \in \mathscr{S}$$

In (**NLP**), R^A and V^A are defined similarly as their counterparts for the defender. It can be shown that, if (V^D, V^A, π^D, π^A) is a global solution to the above optimization problem and the corresponding value of the objective function is zero, then (π^D, π^A) forms an NE of the SG [12].

Value Iteration When the SG admits the zero-sum condition, i.e., when

$$R^A(s, a^D, a^A) = -R^D(s, a^D, a^A), \tag{1.21}$$

the value iteration method can be employed to solve the SG. Particularly, the defender can start with some initial values of the Q- and the value functions and execute the following iterations till convergence:

$$Q_{n+1}^D(s, a^D, a^A) = R^D(s, a^D, a^A) + \beta \cdot \sum_{s'} \mathbb{P}(s'|s, a^D, a^A) V_n^D(s'), \ \forall s, a^D, a^A, \tag{1.22}$$

and

$$V_{n+1}^D(s) \triangleq \max_{\pi^D} \min_{a^A} \sum_{a^D} \pi^D(s, a^D) Q_n(s, a^D, a^A), \ \forall s. \tag{1.23}$$

It can be shown that Q_n^D and V_n^D in the above value iteration method eventually converge to the optimal Q- and value functions, respectively. Accordingly, the NE strategy of the defender is given by

$$\pi_{n+1}^D(s) \triangleq \arg \max_{\pi^D} \min_{a^A} \sum_{a^D} \pi^D(s, a^D) Q_n(s, a^D, a^A), \ \forall s. \tag{1.24}$$

Additionally, it is worth mentioning that, although nothing prevents us from applying the value iteration method to general-sum SGs, the convergence is not guaranteed for general-sum SGs.

In many practical security problems, the state transition probabilities and the reward functions may not be known to the players beforehand. Considering this, various multi-agent reinforcement learning (RL) algorithms have been developed in literature. Roughly speaking, the multi-agent RL algorithms can be treated as some non-trivial extensions of the Q-learning algorithm introduced in the previous subsection. Several widely adopted ones will be introduced in the sequel, and details of multi-agent RL algorithms may be found in, for example, [16] and the references therein.

Minimax-Q The minimax-Q algorithm [15] aims to address the zero-sum SGs and proceeds as follows. At each timeslot n, the defender uses the following equations to up date its Q- and value functions based on the received reward r_n^D and observed state transition from s_n to s_{n+1}:

$$Q_{n+1}^D(s, a^D, a^A) = \begin{cases} (1 - \alpha_n)Q_n^D(s, a^D, a^A) + \alpha_n[r_n^D + \beta V_n^D(s_{n+1})], \\ \qquad\qquad\qquad\qquad \text{for } (s, a^D, a^A) = (s_n, a_n^D, a_n^A), \quad (1.25) \\ Q_n^D(s, a^D, a^A), \qquad\qquad\qquad\qquad\qquad\quad \text{otherwise}, \end{cases}$$

$$V_{n+1}^D(s) = \max_{\pi^D} \min_{a^A} \sum_{a^D} Q_{n+1}^D(s, a^D, a^A)\pi(s, a^D). \qquad (1.26)$$

Based on the updated Q-function, the updated strategy in minimax-Q is given by

$$\pi_{n+1}^D(s) = \arg \max_{\pi^D} \min_{a^A} \sum_{a^D} Q_{n+1}^D(s, a^D, a^A)\pi(s, a^D). \qquad (1.27)$$

It can be shown that Q_{n+1}^D and V_{n+1}^D converge to the optimal Q- and value functions for zero-sum SGs and that the corresponding strategy π_{n+1}^n converges to the NE. Nonetheless, the minimax-Q algorithm is irrational. To better convey the irrationality issue of minimax-Q, several relevant concepts are introduced below [16, 19].

Definition 2 A policy is called *stationary* if it does not change over time (e.g., always taking actions uniformly at random).

Definition 3 A *best response* of an agent is a strategy that achieves the maximum possible expected reward for a given opponent's strategy.

Definition 4 A RL is said to be *rational* if it converges to the best response when the opponent adopts a stationary strategy.

In practice, an irrational learning algorithm may suffer from severe performance loss due to its overly conservative nature. This may be demonstrated through an example given in [19]: Suppose that the second player in a rock-paper-scissors game plays Rock with a very high probability, but Paper and Scissors with some small probabilities. Minimax-Q will lead the first player to a strategy that plays three actions with equal probability; clearly, this is not the best-response since playing only Paper is the only best-response for the considered scenario.

Nash-Q The Nash-Q algorithm [20] is a generalization of the minimax-Q algorithm for non-zero-sum SGs. As compared to the minimax-Q algorithm, the Nash-Q algorithm further requires that each player observes the opponent's reward so as to build a pair of auxiliary Q- and value functions. These auxiliary functions serve as estimates of their counterparts maintained by the opponent. In particular, in the Nash-Q algorithm, the defender updates its own Q-function similarly as in the minimax-Q

$$Q_{n+1}^D(s, a^D, a^A) = \begin{cases} (1 - \alpha_n) Q_n^D(s, a^D, a^A) + \alpha_n [r_n^D + \beta V_n^D(s_{n+1})], \\ \qquad\qquad\qquad \text{for } (s, a^D, a^A) = (s_n, a_n^D, a_n^A), \quad (1.28) \\ Q_n^D(s, a^D, a^A), \qquad\qquad\qquad\qquad\qquad \text{otherwise,} \end{cases}$$

and in addition, it updates the auxiliary Q-function $\tilde{Q}^A$

$$\tilde{Q}_{n+1}^A(s, a^D, a^A) = \begin{cases} (1 - \alpha_n) \tilde{Q}_n^A(s, a^D, a^A) + \alpha_n [r_n^A + \beta \tilde{V}_n^A(s_{n+1})], \\ \qquad\qquad\qquad \text{for } (s, a^D, a^A) = (s_n, a_n^D, a_n^A), \quad (1.29) \\ \tilde{Q}_n^A(s, a^D, a^A), \qquad\qquad\qquad\qquad\qquad \text{otherwise,} \end{cases}$$

where the auxiliary value function is updated by

$$\tilde{V}_{n+1}^A(s) = \text{Nash}^A(Q_{n+1}^D(s, \cdot, \cdot), \tilde{Q}_{n+1}^A(s, \cdot, \cdot)). \tag{1.30}$$

In (4.24), the operator Nash^A finds the value of the attacker in a matrix game with payoff matrices $Q_{n+1}^D(s, \cdot, \cdot)$ and $\tilde{Q}_{n+1}^A(s, \cdot, \cdot)$. The value function of the defender is updated by

$$V_{n+1}^D(s) = \text{Nash}^D\left(Q_{n+1}^D(s, \cdot, \cdot), \tilde{Q}_{n+1}^A(s, \cdot, \cdot)\right). \tag{1.31}$$

Although the Nash-Q algorithm reviewed above is applicable to general-sum SGs, its convergence in only known under very demanding technical conditions that can rarely be met by practical problems.

Win-or-Learn-Fast The essential idea of the WoLF algorithm is to update the strategy with a slow learning rate δ_{win} when the received reward exceeds the player's expectation (i.e., wining), and otherwise, update the strategy with a fast learning rate δ_{lose}. Different from the minimax-Q and the Nash-Q algorithms introduced above, the optimal Q- and value functions of the WoLF algorithm are defined as

$$Q_*^D(s, a^D) \triangleq \mathbb{E}\left[R(s, a^D, a^A) + \beta \cdot V_*^D(s')\right], \tag{1.32}$$

$$V_*^D(s) \triangleq \max_{a^D} Q_*^D(s, a^D). \tag{1.33}$$

It can be seen that the above definitions of the Q- and the value functions are very similar to those used in the Q-learning algorithm discussed in the previous subsection; the only difference is that, on the right-hand side of (1.32), $R(s, a^D, a^A)$ is used instead of $R(s, a)$. Since the defender is always aware of its own reward $r^D = R(s, a^D, a^A)$, there is no need for the defender to directly observe the opponent's action in the WoLF algorithm.

In the WoLF algorithm, to obtain the optimal Q- and value functions, the defender can repeatedly execute the following iterations:

$$
Q_{n+1}^D(s, a^D) = \begin{cases} (1 - \alpha_n)Q_n^D(s, a^D) + \alpha_n[r_n^D + \beta V_n^D(s_{n+1})], \\ \qquad\qquad\qquad\qquad \text{for } (s, a^D) = (s_n, a_n^D), \\ Q_n^D(s, a^D), \qquad\qquad\qquad \text{otherwise}, \end{cases}
$$

and

$$
V_{n+1}^D(s) = \max_{a^D} Q_{n+1}^D(s, a^D). \tag{1.34}
$$

Unlike the minimax-Q and the Nash-Q algorithms that directly derive the updated strategy based on the updated Q-function, the strategy update in the WoLF algorithm depends on whether or not the defender is "winning". Particularly, a state occurrence counter, denoted by c, is maintained by the WoLF algorithm, and $c_n(s)$ represents the number of times that state s has occurred up to timeslot n; it is updated by

$$
c_{n+1}(s) = c_n(s) + \mathbf{1}_{\{s=s_n\}}, \quad \forall s \in \mathscr{S}. \tag{1.35}
$$

Based on the counter, the WoLF algorithm keeps on updating an empirical average strategy $\bar{\pi}$ as follows

$$
\bar{\pi}_{n+1}(s_n, a^D) = \bar{\pi}_n(s_n, a^D) + \frac{\pi_n^D(s_n, a^D) - \bar{\pi}_n(s_n, a^D)}{c(s_n)}, \quad \forall a^D, \tag{1.36}
$$

where π^D is the real strategy taken by the defender at each timeslot and is updated by

$$
\pi_{n+1}^D(s_n, a^D) = \pi_n^D(s_n, a^D) + \Delta_{s_n, a^D}, \quad \forall a^D, \tag{1.37}
$$

In (1.37), the parameter Δ_{s, a^D} is given by

$$
\Delta_{s, a^D} = \begin{cases} -\delta_{s, a^D}, & \text{if } a^D \neq \arg\max_{a'} Q^D(s, a'), \\ \sum_{a' \neq a^D} \delta_{s, a'}, & \text{otherwise}, \end{cases} \tag{1.38}
$$

where

$$\delta_{s,a^D} = \min \left\{ \pi^D(s, a^D), \frac{\delta}{|\mathscr{A}^D| - 1} \right\}. \tag{1.39}$$

The step size δ at each timeslot is determined by the win-or-learn fast principle. Particularly, the defender feels it is winning if the average reward of the current strategy $\sum_{a^D} \pi_n^D(s_n, a^D) Q_{n+1}^D(s_n, a^D)$ is better than that of the average strategy $\sum_{a^D} \bar{\pi}_n(s_n, a^D) Q_{n+1}^D(s_n, a^D)$, and δ is given by

$$\delta = \begin{cases} \delta_{win}, & \text{if } \sum_{a^D} \pi_n^D(s_n, a^D) Q_{n+1}^D(s_n, a^D) > \sum_{a^D} \bar{\pi}_n(s_n, a^D) Q_{n+1}^D(s_n, a^D), \\ \delta_{lose}, & \text{otherwise,} \end{cases}$$

where δ_{win} and δ_{lose} are the step sizes used in the winning and losing cases, respectively, and they both vanish over time. It is worth mentioning that, although the WoLF algorithm does not admit the convergence property in general, it is a rational learning algorithm that can provide the best possible performance when the attacker takes a stationary strategy [16].

Although the detailed analysis for the above learning algorithms will not be provided in this book, we will finish this subsection by introducing an important analytic result that lies the in heart of the analysis of many multi-agent RL algorithms [21].

Theorem 1 *Suppose that the learning rate α_n in a multi-agent RL admits conditions (1.10), and that there exists a sequence of (random) mappings $\mathscr{T}_n : \mathbb{Q} \to \mathbb{Q}$ (with $\mathbb{Q}$ denoting the set of Q-functions) that satisfies:*

(i) $\mathbb{E}[\mathscr{T}_n Q_*] = Q_*$;
(ii) There exists a sequence $\lambda_n > 0$, such that

$$\mathbb{P}\left(\lim_n \lambda_n = 0 \right) = 1, \tag{1.40}$$

and

$$\|\mathscr{T}_n Q - \mathscr{T}_n Q_*\| \le \gamma \cdot \|Q - Q_*\| + \lambda_n, \tag{1.41}$$

where γ is strictly between zero and one, for all $Q \in \mathbb{Q}$. Then, Q_n generated by the following iteration

$$Q_{n+1}(s, a, o) = \begin{cases} (1 - \alpha_n) Q_n(s, a, o) + \alpha_n [(\mathscr{T}_n Q_n)(s, a, o)], \\ \qquad\qquad\qquad\qquad \text{if } (s, a, o) = (s_n, a_n, o_n), \\ Q_n(s, a, o), \qquad\qquad\qquad\qquad \text{otherwise,} \end{cases}$$

converges to Q_* with probability 1.

1.5 Summary

In this chapter, we have reviewed relevant backgrounds to smooth our discussions in later chapters. Specifically, we have reviewed basic concepts in game theory to understand the mathematical framework of analyzing the strategies interactions between the defender and the attacker. In addition, we have conveyed the idea of foresighted optimization for strategic planning in dynamic environments through the framework of MDP. Combining these two fields of theories leads us to the SG framework, which is key to the solutions of many practical security problems. The underlying model and solutions methods of SG have been reviewed in details. In the next chapter, applications of the SG in dynamic security games will be presented.

References

1. D. Fudenberg, J. Tirole, Game theory. MIT Press Books 1(7), 841–846 (2009)
2. M.J. Osborne, A. Rubinstein, *A Course in Game Theory* (MIT, Cambridge, 1994)
3. F. Meshkati, H.V. Poor, S.C. Schwartz, Energy-efficient resource allocation in wireless networks. IEEE Signal Process. Mag. 24(3), 58–68 (2007)
4. J. Pita, M. Jain, F. Ordóñez, C. Portway, M. Tambe, C. Western, P. Paruchuri, S. Krauo, Using game theory for Los Angeles airport security. AI Mag. 30(1), 43 (2009)
5. X. He, H. Dai, P. Ning, R. Dutta, Dynamic IDS configuration in the presence of intruder type uncertainty, in *2015 IEEE Global Communications Conference (GLOBECOM)* (IEEE, New York, 2015), pp. 1–6
6. J.F. Nash, Equilibrium points in n-person games. Proc. Natl. Acad. Sci. 36(1), 48–49 (1950)
7. J. Nash, Non-cooperative games. Ann. Math. 54, 286–295 (1951)
8. I.L. Glicksberg, A further generalization of the Kakutani fixed point theorem, with application to Nash equilibrium points. Proc. Am. Math. Soc. 3(1), 170–174 (1952)
9. J.B. Rosen, Existence and uniqueness of equilibrium points for concave n-person games. Econometrica 33(3), 520–534 (1965)
10. R.A. Howard, Dynamic programming and Markov processes. Math. Gaz. 46(358), 120 (1960)
11. M. Sniedovich, A new look at Bellman's principle of optimality. J. Optim. Theory Appl. 49(1), 161–176 (1986)
12. J. Filar, K. Vrieze, *Competitive Markov Decision Processes* (Springer, Berlin, 1996)
13. C. Watkins, P. Dayan, Q-learning. Mach. Learn. 8(3–4), 279–292 (1992)
14. V. Nunen, A set of successive approximation methods for discounted Markovian decision problems. Z. Oper. Res. 20(5), 203–208 (1976)
15. M.L. Littman, Markov games as a framework for multi-agent reinforcement learning, in *Proceedings of the Eleventh International Conference on Machine Learning*, vol. 157 (1994)
16. L. Busoniu, R. Babuska, B. De Schutter, A comprehensive survey of multiagent reinforcement learning. IEEE Trans. Syst. Man Cybern. C 38(2), 156–172 (2008)
17. T.E.S. Raghavan, J.A. Filar, Algorithms for stochastic games – a survey. Z. Oper. Res. 35(6), 437–472 (1991)
18. Y. Shoham, R. Powers, T. Grenager, Multi-agent reinforcement learning: a critical survey. Web Manuscript (2003)
19. M. Bowling, M. Veloso, Multiagent learning using a variable learning rate. Artif. Intell. 136(2), 215–250 (2002)

20. J. Hu, M.P. Wellman,   Multiagent reinforcement learning: theoretical framework and an algorithm, in *Proceedings of the Fifteenth International Conference on Machine Learning*, vol. 242 (1998), p. 250
21. C. Szepesvári, M. Littman, A unified analysis of value-function-based reinforcement-learning algorithms. Neural Comput. **11**(8), 2017–2060 (1999)

Chapter 2
Overview of Dynamic Network Security Games

2.1 Introduction

Due to environmental variations and system fluctuations, the defender often faces a dynamic security competition against the attacker in practice, and two common challenges exist in addressing these dynamic security games: (1) how to deal with the intelligent attacker that may change its strategy based on the deployed defense, and (2) how to properly align the defense strategy with the environmental or system dynamics to achieve the most efficient and effective defense. In literature, the SG framework reviewed in the previous chapter has been considered as a promising mathematical tool to jointly overcome these two challenges and guide the defender towards the best possible defense.

In this chapter, a brief survey of the state-of-the-art in this direction is provided. Firstly, we will review the application of SG in addressing cyber network security problems such as security and dependability evaluation, intrusion detection, intrusion detection system (IDS) configuration, cyber attack prediction, insider attack, and interdependent network security. Then, our discussion will be devoted to SG based solutions to jamming problems in wireless networks. Lastly, based on relevant existing literature, we will discuss how SG can be applied to address various security issues in cyber-physical systems, including power grid protection, plant protection, anti-jamming in networked control systems, and risk minimization in automatic generation control.

© The Author(s) 2018
X. He, H. Dai, *Dynamic Games for Network Security*, SpringerBriefs in Electrical and Computer Engineering, https://doi.org/10.1007/978-3-319-75871-8_2

2.2 Applications in Cyber Networks

This section aims to provide a brief survey of the applications of SG in addressing various security problems in cyber networks.

Firstly, the SG reviewed in the previous chapter can be employed to address the *security and dependability evaluation problem* for the domain name system (DNS). For example, in [1], a DNS security problem with three possible attacks including the illegal login attack, the cache poisoning attack, and the server shut down attack is considered. In this work, depending on the actions taken by the defender and the attacker, the DNS is assumed to stochastically transit among six different states (e.g., software integrity failure state, hardware failure state, etc.), and then, under a zero-sum assumption, the value iteration algorithm is employed to solve this stochastic security game so as to predict the most likely strategy of the attacker and design the corresponding optimal defense. A similar application was considered in [2], where the equilibrium strategies of the network administrator and the attacker are computed using the non-linear programming approach [3].

The SG has also been adopted to analyze the *intrusion detection problem*. For example, in [4], the interactions between an intruder and an IDS is modeled as a zero-sum SG. The action space of the intruder constitutes of all the possible attacks from the intruder while that of the IDS includes different defense actions (e.g., setting an alert and gathering further information). The state of this dynamic security game is defined as the output of an auxiliary sensor network (e.g., a certain attack is detected or the target system is safe), which is influenced by the state variation of the underlying target system as well as the actions taken by both the intruder and the IDS. The minimax-Q algorithm is employed to solve this security game and find the optimal defense strategy of the IDS. Real-world experiments of using the SG to address a similarly structured problem were conducted in [5]. The Stackelberg version of the SG has also been considered in literature to model the interaction between the intruder and a detection and response system [6]. In addition, due to limited resource, practical IDS usually can only activate a limited number of detection algorithms at a time and hence choosing a proper set of detection algorithms is of paramount importance for intrusion detection. Such *IDS configuration problem* is examined in [7] through the lens of SG. In the corresponding SG, the IDS can choose different configurations as its actions to defend against various possible attacks from the attacker. State transition in the corresponding SG captures the state changes in the computer system protected by the IDS. The reward of the defender considered there captures both the configuration cost and the potential damage incurred by undetected attacks. Both Newton's method [3] and the value iteration method are employed to find the values of this SG and the corresponding optimal IDS configuration. To handle the cases of misaligned objective functions between the defender and the attacker, this work was extended to a non-zero sum stochastic security game in [8], and non-linear programming was adopted to find the corresponding equilibrium strategy.

With the objective of *cyber attacker prediction*, the SG has been employed in [9] to analyze the security competition between the defender and the attacker over a data fusion system. Particularly, to capture the effect of cooperative defense and large-scale attack, the payoff functions of the SG considered there includes both the rewards generated by the mutual cooperation and that induced by the opponents' action. Also, instead of considering the discounted rewards accumulated over an infinite-horizon as most of the existing literature, [9] only intends to optimize the reward over a k time-step horizon by leveraging the fictitious play algorithm [10].

The SG has also been utilized to facilitate the design of defense mechanisms against the *insider attacks* [11]. Specifically, by modeling the computing system as a state machine, the state of the SG can be defined as the security status of the computing system, which changes over time in a probabilistic manner depending on the actions taken by the defender and the insider. Exemplary actions of the defender and the insider considered in [11] include setting up bogus information and revoking a user's privilege, respectively. To simplify the analysis, a zero-sum assumption was taken when modeling the payoff functions of both parties and the nonlinear programming approach [3] is adopted to find the equilibrium of this SG so as to predict the most possible attacking strategy of the insider.

Another application of SG in cyber networks is to address *security issues in interdependent networks* [12]. In such networks, due to the interdependence among the network nodes, once the security status of a single node is affected by the actions of the defender and the attacker, a security status change of all the related nodes may be triggered and hence generates complex system dynamics. For this reason, SG has been considered as a suitable tool for analyzing the corresponding security competitions and predicting the possible behaviors of the attacker. The existence, uniqueness and structure of the equilibrium strategies of the defender and the attacker are analyzed in [12]. By combining SG and Petri nets, the authors in [13] proposed an SG net to analyze the security issues in enterprise computer networks, and the nonlinear programming approach is employed to find the equilibrium strategies.

2.3 Applications in Wireless Networks

The SG can also serve as a powerful analytic tool for addressing security problems in wireless networks. For example, based on the SG framework, the *anti-jamming* problem is investigated by Chen et al. [14] in the context of multi-channel cognitive radio systems. Particularly, the secondary user's objective is to select the right channel and transmit power so as to deliver more data with the minimum effort, whereas the jammer aims at exactly the opposite, leading to a zero-sum security game. To confine the size of the action space, it is assumed in [14] that finite numbers of transmitting and jamming power levels are available to the secondary user and the jammer, respectively, and that both of them can only access one of the multiple channels at each timeslot. In addition, the actions taken by the secondary user and the jammer at the previous timeslot are defined as the state of the underlying stochastic security game; consequently, the state transition is controlled by the players' strategies. The WoLF algorithm was employed to find the best channel selection and transmit power control strategy for the secondary user. A similar jamming and anti-jamming competition is considered in [15] but the focus is on the control channel instead of the data channel. To reduce the computational load of solving such stochastic security game, the secondary user can employ the conventional Q-learning algorithm to obtain a heuristic solution with suboptimal performance [16]. Unlike [14] and [16], [17] considers a stochastic anti-jamming game where the defender aims to optimize its average payoff rather than the discounted one. The anti-jamming problem in cognitive radio networks has also been considered in [18] where both the control and the data channels are included in decision-making and minimax-Q is adopted to find the optimal defense. In addition to the cases of a single pair of defender and attacker, the jamming and anti-jamming problem between multiple collaborative defenders (i.e., the blue force) and colluding attackers (i.e., the red force) is considered [19]. Two types of nodes are assumed for each force: the communication nodes and the jamming nodes. Each force aims to leverage its own communication nodes to achieve the best possible data transmission and the jamming nodes to suppress the opponent's communication. An action of this security game is an assignment of these nodes to (a subset of) the channels. The state of this dynamic security game is defined as the channel assignments of both the red and the blue forces in the previous timeslots. The reward of each force consists of two parts; one part is the reward due to successful data delivery by its own communication nodes while the other part is due to successful jamming of the opponent's transmission. In [19], the performances of three different reinforcement learning algorithms are compared for this stochastic security game, including the minimax-Q algorithm, Nash-Q algorithm, and the Friend-or-Foe-Q algorithm [20]. The corresponding results suggest that, for the considered application, minimax-Q algorithm is suitable for aggressive environments and the Friend-or-Foe-Q algorithm is more suitable for distributed settings.

2.4 Applications in Cyber-Physical Networks

SGs have also been widely considered to address the security problems in cyber-physical systems and networks, and some exemplary applications will be reviewed below.

Power grid as one of the largest cyber-physical systems is usually the target of many adversarial activities. To better *protect a power grid* with the often limited security resource, SG has been employed in [21] to model and analyze the most likely attacking strategy and the corresponding best possible defense. Particularly, the defender of the power grid has to strategically enforce a subset of transmission lines (e.g., preparing more personnel and materials for repair in the event of an attack) and repair the broken lines with its limited security resource. While the attacker aims to attack the transmission lines that can cause the most disruption. The state of the SG is described by the up/down status of the transmission lines in the power grid, and the payoff of the players are measured by functions of the total amount of shed load. By assuming known state transition probabilities, value iteration is employed to find the equilibrium strategy of the game.

The *security competition between the controller of a plant and a replay attacker* is modeled as a stochastic security game in [22]. Particularly, the replay attacker is assumed to be able to intercept all the sensor outputs from the plant and feed a delayed version of the sensor measurements to the monitoring system (i.e., a state estimator). The corresponding action space of the attacker consists of all the possible $(m + 1)$ delayed versions of the measurement, with m the maximum possible delay. To facilitate the detection of such replay attack, the controller can embed some authentication signal into the control signal, which however, may decrease the control performance. That being said, the controller has to properly chosen between two actions—to embed, or not to embed. Three system states are considered for the plant. The first state is a safe state which indicates that the system has already successfully caught the attacker; the second state is a no-detection state, meaning that the system has not yet detected any attack, the last state is a false alarm state which happens when a false alarm is generated. The reward functions of the controller and the replay attacker are determined by the linear-quadratic-Gaussian (LQG) cost. To account for the possible uncertainties in relevant parameters, the robust game theory [23] is incorporated into the value iteration algorithm to solve the corresponding SG [22]. A similar security competition over a resilience control system was considered in [24]. In the corresponding stochastic security game, the payoff function includes both the cost inflicted on the cyber layer of the system and the influence on the physical layer control system.

The *anti-jamming issue in networked control systems* has been investigated in [25]. In this work, the following linear state feedback system is considered: $x_{n+1} = Ax_n + Bu_n$, and $u_n = Kx_n$, where A and B are the state and input matrices, respectively, and K is the feedback matrix. The objective of the jammer is to prevent the sensor from sending the state information to the controller, whereas the defender's objective is to raise the transmission power of the sensor to overcome jamming; it is assumed that, when jamming is successful, the controller will erroneously observe $x_n = \mathbf{0}$. The state $s_n = (x_n, \gamma_n)$ of this SG at each timeslot n consists of two parts where the first part is the (discretized) state of the control system and the second part $\gamma_n \in \{0, 1\}$ indicates whether or not jamming is successful in the previous timeslot. At each timeslot, the defender and the jammer choose an appropriate transmitting power a^D and jamming power a^J, respectively. The resulting cost to the defender includes both the power consumption and the performance degradation of the system; whereas the jammer's cost is the opposite due to a zero-sum assumption. By assuming the knowledge of relevant system parameters, the linear programming approach is used to find the optimal defense strategy.

Risk minimization in automatic generation control was studied in [26] by leveraging the stochastic security game framework. In this problem, the adversary intends to inject falsified frequency measurements into the generator's sensor so as to trigger load shedding and inflict revenue loss on the energy provider. The underlying target system is an interconnected system with two control areas, and the corresponding system state is determined by the frequency deviations in both areas. The action space of the attacker constitutes of four different attacks: (1) constant injection, in which the attacker injects a constant falsified frequency deviation Δf, (2) basic injection, in which the attacker adds a falsified Δf on top of the true frequency deviation, (3) overcompensation, in which the attacker amplifies the true frequency deviation k times with k a large positive number, and (4) negative compensation, which is similar to overcompensation but with a negative k. As to the defender, three basic defense mechanisms are assumed: (1) saturation filter, in which the input to sensor is restricted to a certain range, (2) redundancy, in which multiple frequency meters are activated to ensure that the defender can obtain the genuine frequency measurement with high probability, and (3) detection, in which certain anomaly detection algorithms will be activated. The cost of the defender covers not only the influence on the revenue but also the impact of false alarm (when detection algorithm is activated). Value iteration is used to solve this dynamic security game.

2.5 Summary

In this chapter, we have reviewed existing applications of stochastic security game, in traditional cyber networks where the dynamics are mainly triggered by network node status change, in wireless communication networks where the dynamics are caused by the variation of wireless channel, and in the recently emerged cyber-physical systems where the dynamics are due to state change in the physical systems. Nonetheless, most of these existing works focus on scenarios where the defender and the attacker hold equal information of the ongoing security competitions. In the next chapter, we will review some state-of-the-art on SG with information asymmetry.

References

1. K. Sallhammar, B.E. Helvik, S.J. Knapskog, On stochastic modeling for integrated security and dependability evaluation. J. Networks **1**(5), 31–42 (2006)
2. K.W. Lye, J.M. Wing, Game strategies in network security. Int. J. Inf. Secur. **4**(1), 71–86 (2005)
3. J. Filar, K. Vrieze, *Competitive Markov Decision Processes* (Springer Science & Business Media, Berlin, 2012)
4. T. Alpcan, T. Basar, A game theoretic analysis of intrusion detection in access control systems, in *IEEE Conference on Decision and Control*, vol. 2 (IEEE, New York, 2004), pp. 1568–1573

5. W. Jiang, Z. Tian, H. Zhang, X. Song, A stochastic game theoretic approach to attack prediction and optimal active defense strategy decision, in *IEEE International Conference on Networking, Sensing and Control, 2008 (ICNSC 2008)* (IEEE, New York, 2008), pp. 648–653
6. S.A. Zonouz, H. Khurana, W.H. Sanders, T.M. Yardley, RRE: a game-theoretic intrusion response and recovery engine. IEEE Trans. Parallel Distrib. Syst. **25**(2), 395–406 (2014)
7. Q. Zhu, T. Başar, Dynamic policy-based IDS configuration, in *IEEE Conference on Decision and Control* (IEEE, New York, 2009), pp. 8600–8605
8. Q. Zhu, H. Tembine, T. Basar, Network security configurations: a nonzero-sum stochastic game approach. **20**(1), 1059–1064 (2010)
9. D. Shen, G. Chen, E. Blasch, G. Tadda, Adaptive Markov game theoretic data fusion approach for cyber network defense, in *IEEE Military Communications Conference* (IEEE, New York, 2007), pp. 1–7
10. D. Fudenberg, D.K. Levine, *The Theory of Learning in Games*, vol. 2 (MIT, Cambridge, 1998)
11. D. Liu, X. Wang, J. Camp, Game-theoretic modeling and analysis of insider threats. Int. J. Crit. Infrastruct. Prot. **1**, 75–80 (2008)
12. K.C. Nguyen, T. Alpcan, T. Basar, Stochastic games for security in networks with interdependent nodes, in *ICST International Conference on Game Theory for Networks* (2009), pp. 697–703
13. C. Lin, Y. Wang, Y. Wang, H. Zhu, Q.L. Li, Stochastic game nets and applications in network security. J. Comp. (2011), pp. 461–467
14. T. Chen, J. Liu, L. Xiao, L. Huang, Anti-jamming transmissions with learning in heterogenous cognitive radio networks, in *2015 IEEE Wireless Communications and Networking Conference Workshops (WCNCW)* (IEEE, New York, 2015), pp. 293–298
15. B.F. Lo, I.F. Akyildiz, Multiagent jamming resilient control channel game for cognitive radio ad hoc networks, in *2012 IEEE International Conference on Communications (ICC)* (IEEE, New York, 2012), pp. 1821–1826
16. C. Chen, M. Song, C. Xin, J. Backens, A game-theoretical anti-jamming scheme for cognitive radio networks. IEEE Netw. **27**(3), 22–27 (2013)
17. Q. Zhu, H. Li, Z. Han, T. Basar, A stochastic game model for jamming in multi-channel cognitive radio systems, in *2010 IEEE International Conference on Communications (ICC)* (IEEE, New York, 2010), pp. 1–6
18. B. Wang, Y. Wu, K.J.R. Liu, T.C. Clancy, An anti-jamming stochastic game for cognitive radio networks. IEEE J. Sel. Areas Commun. **29**(4), 877–889 (2011)
19. Y. Gwon, S. Dastangoo, C. Fossa, H. Kung, Competing mobile network game: embracing antijamming and jamming strategies with reinforcement learning, in *2013 IEEE Conference on Communications and Network Security (CNS)* (IEEE, New York, 2013), pp. 28–36
20. M.L. Littman, Friend-or-foe Q-learning in general-sum games, in *ICML*, vol. 1 (2001), pp. 322–328
21. C. Ma, D. Yau, X. Lou, N. Rao, Markov game analysis for attack-defense of power networks under possible misinformation. IEEE Trans. Power Syst. **28**(2), 1676–1686 (2013)
22. F. Miao, M. Pajic, G.J. Pappas, Stochastic game approach for replay attack detection, in *2013 IEEE 52nd Annual Conference on Decision and Control (CDC)* (IEEE, New York, 2013), pp. 1854–1859
23. M. Aghassi, D. Bertsimas, Robust game theory. Math. Program. **107**(1), 231–273 (2006)
24. Q. Zhu, T. Basar, Game-theoretic methods for robustness, security, and resilience of cyberphysical control systems: games-in-games principle for optimal cross-layer resilient control systems. IEEE Control Syst. **35**(1), 46–65 (2015)
25. S. Liu, P.X. Liu, A. El Saddik, A stochastic game approach to the security issue of networked control systems under jamming attacks. J. Frankl. Inst. **351**(9), 4570–4583 (2014)
26. Y.W. Law, T. Alpcan, M. Palaniswami, Security games for risk minimization in automatic generation control. IEEE Trans. Power Syst. **30**(1), 223–232 (2015)

Chapter 3
Dynamic Security Games with Extra Information

3.1 Introduction

In the previous chapter, we have reviewed existing applications of SG in addressing different dynamic security problems. Nonetheless, most existing works assume equal knowledgeable defender and attacker, whereas, in practice, the defender and the attacker are likely to hold different information about the ongoing security rivalries. For example, in an energy harvesting communication system (EHCS), the energy harvesting rate is often known to the legitimate system but not the jammer. As another example, in a cloud-based security system, the statistical property of the available security resource in the cloud may only be known to the defender but not the adversary. Conventional SG based methods developed in existing literature do not fit well to such dynamic security games with information asymmetry. To achieve the best possible defense in such scenarios, new techniques that can adequately manage the information asymmetry are needed.

In this chapter, we will focus on scenarios where the defender has extra information and present two novel algorithms, termed minimax-PDS and WoLF-PDS, to solve the corresponding dynamic security games with extra information. These two algorithms are motivated by the single-agent post-decision state (PDS)-learning that enables the agent to expedite its learning speed in MDP by properly exploiting its extra information [1]. The essential idea of the PDS is to properly define a PDS so as to allow the agent to simultaneously update multiple Q-functions (instead of a single one as in conventional Q-learning) for faster learning. For this reason, we will first illustrate the basic idea of PDS-learning in the single-agent setting. Then, we proceed to demonstrate how to incorporate the PDS concept into the SG framework and extend the minimax-Q and WoLF algorithms to minimax-PDS and WoLF-PDS, respectively.

© The Author(s) 2018
X. He, H. Dai, *Dynamic Games for Network Security*, SpringerBriefs in Electrical
and Computer Engineering, https://doi.org/10.1007/978-3-319-75871-8_3

After establishing the theoretic basis, we then provide two exemplary applications to illustrate how these two new algorithms can be leveraged to handle dynamic security games with extra information. The first application is concerned with anti-jamming in EHCS. Due to the ever-increasing demand for a larger volume of data exchange, traditional battery powered wireless devices easily get depleted. Recently, by taking advantage of the advancement in energy harvesting techniques, EHCS has emerged as a promising solution to this problem [2, 3]. Nevertheless, as in conventional wireless systems, jamming always remains as a haunting threat to EHCS. Although abundant number of anti-jamming techniques have been developed in the past decades [4–13], they are not tailor-made for EHCS and hence cannot fully exploit some unique features of the EHCS (e.g., the extra information about the energy harvesting rate). With this consideration, we will demonstrate how the minimax-PDS and the WoLF-PDS algorithms can enable the EHCS to boost its learning and adaptation by exploiting its extra information. The second application is concerned with a cloud-based security game. With the recent advancement in cloud-computing techniques [14], it becomes possible to implement cloud-based defense systems [15, 16] that can leverage the more powerful cloud to perform computationally demanding security mechanisms [17, 18] for a better defense. In practice, the amount of resource available at the cloud may change over time (e.g., due to internal scheduling and variations of customer demands), and such statistical information is often only known to the defender. Naturally, this leads to a dynamic security game with extra information, which again, can be addressed by the minimax-PDS and the WoLF-PDS algorithms.

3.2 Post-decision State

Fulfilling the idea of boosting the defender's adaptation by exploiting its extra information relies critically on the notion of post-decision state (PDS) [1]. In this section, some background of the single-agent PDS is reviewed first to pave the way for the later exhibition of multi-agent PDS and its application in dynamic security games.

When the learning agent holds extra statistical information about the state transition probability and the structure of the reward function, the notion of single-agent PDS can be employed to accelerate learning in MDP (c.f. Chap. 1).

Definition 5 The PDS, often denoted as $\tilde{s}$, is an intermediate state appeared after the agent taking its action a in the current state s but before the occurrence of the future state s'.

Usually, the defined PDS of a certain MDP admits the following property: First, the original state transition probability $p(s'|s, a)$ can be divided into two parts, $p^k(\tilde{s}|s, a)$ (i.e., the extra information) and $p^u(s'|\tilde{s}, a)$, which indicate the *known* probability of transiting from the current state s to the PDS $\tilde{s}$ and the *unknown* probability of transiting from the PDS to the future state s'. Accordingly, the reward function can be split into two parts. The first part $r^k(s, a)$ represents the known reward received before the occurrence of the PDS while the second part $r^u(\tilde{s}, a)$ captures the reward received when the system switches from $\tilde{s}$ to s'; note that the unknown part of the reward does not depend on the current state s. This notion is further illustrated in Fig. 3.1. More specifically, the state transition probabilities and the reward functions admit

$$p(s'|s, a) = \sum_{\tilde{s}} p^u(s'|\tilde{s}, a) p^k(\tilde{s}|s, a), \tag{3.1}$$

and

$$\mathbb{E}_{\tilde{s}}[R(s, a)] = r^k(s, a) + \sum_{\tilde{s}} p^k(\tilde{s}|s, a) r^u(\tilde{s}, a). \tag{3.2}$$

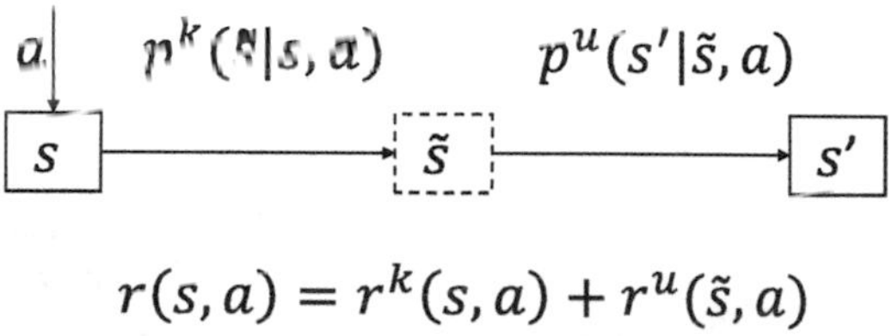

Fig. 3.1 Single-agent PDS

$$r(s, a) = r^k(s, a) + r^u(\tilde{s}, a)$$

Once the PDS is properly defined, it can be used to accelerate the learning speed of the conventional Q-learning, and the resulting method is termed *PDS-learning*. To implement the PDS-learning, a quality function $\tilde{Q}^{(p)}(\tilde{s}, a)$ has to be defined for each PDS-action pair $(\tilde{s}, a)$, and this quality function has a similar interpretation as the Q-function in the conventional Q-learning algorithm. To avoid any potential misunderstanding, a superscript (p) is added to the quality function of the PDS.[1] The PDS quality function is updated at each timeslot by using the following equation

$$\tilde{Q}_{n+1}^{(p)}(\tilde{s}_n, a_n) = (1 - \alpha_n)\tilde{Q}_n^{(p)}(\tilde{s}_n, a_n) + \alpha_n \left[r^u(\tilde{s}_n, a_n) + \beta \cdot V_n^{(p)}(s_{n+1}) \right], \tag{3.3}$$

---

[1] In the rest part of this chapter, superscripts (m), (w), (mp) and (wp) will be used for the quality and value functions in the conventional minimax-Q and WoLF, and the proposed minimax-PDS and WoLF-PDS algorithms, respectively.

where

$$V_n^{(p)}(s) \triangleq \max_a Q_n^{(p)}(s, a). \tag{3.4}$$

After obtaining the updated PDS quality function $\tilde{Q}_{n+1}^{(p)}$, the original Q-function for *all* the state-action pairs (s, a) can be updated as follows:

$$Q_{n+1}^{(p)}(s, a) = r^k(s, a) + \sum_{\tilde{s}} p^k(\tilde{s}|s, a) \tilde{Q}_{n+1}^{(p)}(\tilde{s}, a), \quad \forall s, a, \tag{3.5}$$

and the policy is updated as in the conventional Q-learning algorithm. As illustrated in Fig. 3.2, the above PDS-learning allows the agent to update multiple Q-function according to (3.5) at each timeslot by exploiting the extra information $p^k(\tilde{s}|s, a)$ and $r^k(s, a)$; in contrast, the conventional Q-learning algorithm can only update a single Q-function. Consequently, PDS-learning is usually significantly faster than the conventional Q-learning algorithm.

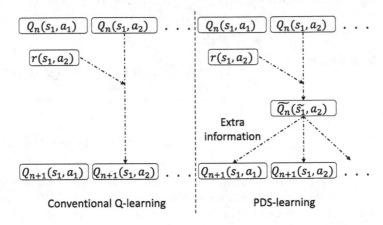

Fig. 3.2 Comparison of the conventional Q-learning and the PDS-learning

3.3 Multi-Agent PDS Learning

Bearing the PDS concept in mind, in this section, we will illustrate how the PDS-learning can be extended to the multi-agent scenarios and used to boosting the learning speed of the agent with extra information. In particular, by incorporating

PDS-learning into the two classic multi-agent reinforcement learning algorithms, minimax-Q and WoLF-Q, we will develop two new multi-agent PDS-learning algorithms, termed minimax-PDS and WoLF-PDS. In the rest of this section, considering that these two algorithms may be applied to a wide range of engineering problems, the discussion will be given in the context of general SG; consequently, we will slightly abuse the notation by using o, instead of a^A as in Chap. 1, to represent the action from the opponent. The security applications of these two algorithms will be discussed in the next section.

Fig. 3.3 Multi-agent
PDS-learning

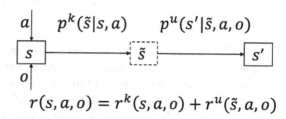

$$r(s, a, o) = r^k(s, a, o) + r^u(\tilde{s}, a, o)$$

For clarity, in the discussion of the rest of this section, the agent with extra information will be called the PDS agent. Similar to the single-agent setting, the extra information available to the PDS agent are the transition probability $p^k(\tilde{s}|s, a)$ from the current state s to the PDS $\tilde{s}$ and the corresponding part of the reward function $r^k(s, a, o)$. It is worth emphasizing that, due to some technical reasons, $p^k(\tilde{s}|s, a)$ is assumed only depending on the action a from the PDS agent. In addition to the known part of the state transition, the unknown part of the state transition probability will be denoted by $p^u(s'|\tilde{s}, a, o)$ and the corresponding part of the reward function is denoted by $r^u(s, a, o)$. The relation among these quantities is illustrated in Fig. 3.3. Mathematically, the PDS and the extra information must admit the following relations

$$p(s'|s, a, o) = \sum_{\tilde{s}} p^u(s'|\tilde{s}, a, o)p^k(\tilde{s}|s, a), \qquad (3.6)$$

and

$$\mathbb{E}_{\tilde{s}}[R(s, a, o)] = r^k(s, a, o) + \sum_{\tilde{s}} p^k(\tilde{s}|s, a)r^u(\tilde{s}, a, o). \qquad (3.7)$$

In the sequel, the minimax-PDS and the WoLF-PDS algorithms will be introduced, respectively.

3.3.1 The Minimax-PDS Algorithm

A similar approach as in the single-agent PDS-learning will be used to extend
the conventional minimax-Q algorithm to minimax-PDS algorithm. Particularly, a
PDS quality function, denoted by $\tilde{Q}^{(mp)}$, is defined in the minimax-PDS algorithm.
Mathematically, the optimal quality function $\tilde{Q}_*^{(mp)}$ for the PDS-action pair $(\tilde{s}, a, o)$
is defined as

$$\tilde{Q}_*^{(mp)}(\tilde{s}, a, o) \triangleq r^u(\tilde{s}, a, o) + \beta \cdot \sum_{s'} p^u(s'|\tilde{s}, a, o) V_*^{(mp)}(s'), \qquad (3.8)$$

where $Q_*^{(mp)}$ and $V_*^{(mp)}$ are the optimal quality and the optimal value functions of
the minimax-PDS algorithm, and they carry similar interpretation as their counterparts in the conventional minimax-Q algorithm. The basic idea of the minimax-PDS
algorithm is to update the PDS quality function $\tilde{Q}^{(mp)}$ first and then simultaneously
update multiple standard Q-functions by exploiting the extra information $p^k(\tilde{s}|s, a)$
and $r^k(s, a, o)$. When the optimal PDS quality function $\tilde{Q}_*^{(mp)}(\tilde{s}, a, o)$ is available,
the standard Q-function $Q_*^{(mp)}(s, a, o)$ can be directly evaluated by leveraging the
extra information. More specifically, we have the following result.

Theorem 2 *Given the optimal PDS quality function* $\tilde{Q}_*^{(mp)}(\tilde{s}, a, o)$ *and the extra
information* $p^k(\tilde{s}|s, a)$ *and* $r^k(s, a, o)$*, the standard Q-function* $Q_*^{(mp)}(s, a, o)$ *for
all state-action pairs* (s, a, o) *are given by*

$$Q_*^{(mp)}(s, a, o) = r^k(s, a, o) + \sum_{\tilde{s}} p^k(\tilde{s}|s, a) \cdot \tilde{Q}_*^{(mp)}(\tilde{s}, a, o), \text{ for all } s, a, o. \quad (3.9)$$

Proof The above result can be readily proved as follows:

$$
\begin{aligned}
&Q_*^{(mp)}(s, a, o) \\
&= \mathbb{E}_{\tilde{s}}[R(s, a, o)] + \beta \sum_{s'} p(s'|s, a, o) V_*^{(mp)}(s') \\
&= r^k(s, a, o) + \sum_{\tilde{s}} p^k(\tilde{s}|s, a) \left[r^u(\tilde{s}, a, o) + \beta \cdot \sum_{s'} p^u(s'|\tilde{s}, a, o) V_*^{(mp)}(s') \right] \\
&= r^k(s, a, o) + \sum_{\tilde{s}} p^k(\tilde{s}|s, a) \tilde{Q}_*^{(mp)}(\tilde{s}, a, o), \qquad\qquad\qquad\qquad (3.10)
\end{aligned}
$$

where the first equality follows from the Bellman equation; the second one follows
from (3.6) and (3.7); the last one follows from the definition in (3.8).

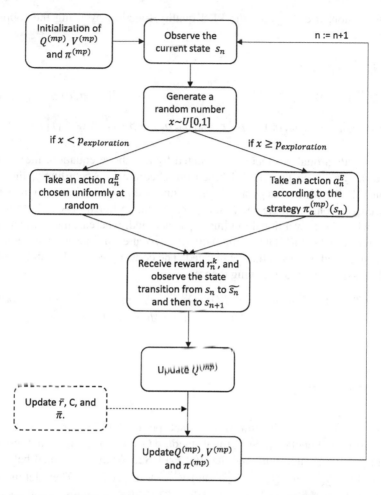

Fig. 3.4 The minimax-PDS algorithm

The overall diagram of the minimax-PDS algorithm is plotted in Fig. 3.4 (excluding the dashed box). At the beginning of the SG, the PDS agent initializes the relevant quantities $Q^{(mp)}$, $V^{(mp)}$ and $\pi^{(mp)}$. At each timeslot, the PDS agent observes the current state s_n and generates a number $x \in [0, 1]$ uniformly at random. If the random number is less than the pre-determined exploration rate $p_{exploration}$, the agent will take an action a_n chosen uniformly at random; otherwise, the action is chosen according to the current strategy $\pi_a^{(mp)}(s_n)$. Then, the PDS agent receives its reward $r^k(s_n, a_n, o_n)$ and $r^u(\tilde{s}_n, a_n, o_n)$, and observes the state transits from the current one s_n to the PDS $\tilde{s}_n$ and then to the next state s_{n+1}. After collecting all this information, it can update the PDS quality function by using the following equation

$$\tilde{Q}_{n+1}^{(mp)}(\tilde{s}_n, a_n, o_n) = \tag{3.11}$$

$$\begin{cases} \tilde{Q}_n^{(mp)}(\tilde{s}, a, o), & \text{if } (\tilde{s}, a, o) \neq (\tilde{s}_n, a_n, o_n), \\ (1 - \alpha_n)\tilde{Q}_n^{(mp)}(\tilde{s}_n, a_n, o_n) + \alpha_n\left[r^u(\tilde{s}_n, a_n, o_n) + \beta \cdot V_n^{(mp)}(s_{n+1})\right], & \text{otherwise.} \end{cases}$$

Once the PDS quality function is updated by the above equation, the standard Q-function $Q^{(mp)}$ can be updated based on Theorem 2. Accordingly, the value function $V^{(mp)}$ and the strategy $\pi^{(mp)}$ can be updated as well. Similar to the single-agent scenarios, only a single entry (s_n, a_n, o_n) of the Q-function is updated at each timeslot when the conventional minimax-Q is adopted; whereas, the minimax-PDS algorithm can update *all* state-action pairs (s, a, o) at each timeslot. Consequently, the learning speed is substantially enhanced. The convergence of the minimax-PDS algorithm is given by the following theorem.

Theorem 3 ([19]) *Using minimax-PDS, $Q_n^{(mp)}$ converges to $Q_*^{(mp)}$ with probability 1, when the learning rate sequence $\{\alpha_n\}_{n \geq 1}$ satisfies $0 \leq \alpha_n < 1$, $\sum_{n=0}^{\infty} \alpha_n = \infty$ and $\sum_{n=0}^{\infty} \alpha_n^2 < \infty$.*

3.3.2 WoLF-PDS

In this subsection, we will illustrate how to incorporate PDS-learning into the classic WoLF algorithm. For clarity, the optimal standard Q-function (value function) of the conventional WoLF algorithm and the WoLF-PDS algorithm presented below will be denoted by $Q_*^{(w)}$ and $Q_*^{(wp)}$ ($V_*^{(w)}$ and $V_*^{(wp)}$), respectively. Then, let us define the PDS quality function $\tilde{Q}_*^{(wp)}(\tilde{s}, a)$ of the WoLF-PDS algorithm as follows

$$\tilde{Q}_*^{(wp)}(\tilde{s}, a) \triangleq \mathbb{E}_o\left[r^u(\tilde{s}, a, o) + \beta \sum_{s'} p^u(s'|\tilde{s}, a, o)V_*^{(wp)}(s')\right]. \tag{3.12}$$

With this definition, a similar result as in the minimax-PDS case can be obtained.

Theorem 4 *Given the optimal PDS quality function $\tilde{Q}_*^{(wp)}(\tilde{s}, a)$ and the extra information $p^k(\tilde{s}|s, a)$ and $r^k(s, a, o)$, the standard Q-function for all the state-action pair (s, a) can be computed as*

$$Q_*^{(wp)}(s, a) \triangleq \mathbb{E}_o[r^k(s, a, o)] + \sum_{\tilde{s}} p^k(\tilde{s}|s, a)\tilde{Q}_*^{(wp)}(\tilde{s}, a). \tag{3.13}$$

Similar method as in Theorem 2 can be employed to prove the above result, and the details can be found in [19].

The implementation of WoLF-PDS is similar to that of the minimax-PDS as depicted in Fig. 3.4, except for the additional update procedure in the dashed box. In particular, at each timeslot, after collecting the relevant information s_n, a_n, o_n, $r^k(s_n, a_n, o_n)$, $\tilde{s}_n$, $r^u(\tilde{s}_n, a_n, o_n)$, and s_{n+1}, the PDS agent can update its PDS quality function by using the following equation

$$\tilde{Q}_{n+1}^{(wp)}(\tilde{s}_n, a_n) = \tag{3.14}$$

$$\begin{cases} \tilde{Q}_n^{(wp)}(\tilde{s}, a), & \text{if } (\tilde{s}, a) \neq (\tilde{s}_n, a_n), \\ (1 - \alpha_n)\tilde{Q}_n^{(wp)}(\tilde{s}_n, a_n) + \alpha_n[r^u(\tilde{s}_n, a_n, o_n) + \beta \cdot V_n^{(wp)}(s_{n+1})], & \text{otherwise.} \end{cases}$$

Different from the minimax-PDS algorithm, a WoLF-PDS agent has to further maintain a record about the empirical average performance of taking action a at state s. To this end, the PDS agent updates an empirical reward function $\bar{r}^k$ at each timeslot for *all* a by using the following equation

$$\bar{r}_{n+1}^k(s_n, a) = \begin{cases} \bar{r}_n^k(s, a), & \text{if } s \neq s_n, \\ (1 - \alpha_n)\bar{r}_n^k(s_n, a) + \alpha_n \cdot r^k(s_n, a, o_n), & \text{otherwise.} \end{cases} \tag{3.15}$$

Once both the updated PDS quality function $\tilde{Q}_{n+1}^{(wp)}$ and the empirical reward function $\bar{r}_{n+1}^k$ are obtained, the standard Q-function $Q_{n+1}^{(wp)}(s, a)$ will be adjusted by

$$Q_{n+1}^{(wp)}(s, a) = \bar{r}_{n+1}^k(s, a) + \sum_{\tilde{s}} p^k(\tilde{s}|s, a)\tilde{Q}_{n+1}^{(wp)}(\tilde{s}, a), \quad \forall s, a. \tag{3.16}$$

The above equation indicates that the WoLF-PDS algorithm can update multiple standard Q-functions $Q_{n+1}^{(wp)}$ at each timeslot and thus substantially expedite the learning speed. The rest steps of WoLF-PDS for updating the state occurrence count C, the empirical average policy $\bar{\pi}$, and the policy $\pi^{(wp)}$ are the same as the original WoLF algorithm discussed in Chap. 1.

Neither the conventional WoLF algorithm nor the WoLF-PDS algorithm has the convergence property as the minimax-PDS algorithm. Nonetheless, the rationality property of the WoLF algorithm is inherited by the WoLF-PDS algorithm. In particular, we have the following result.

Theorem 5 ([19]) *The WoLF-PDS is rational, when the learning rate sequence* $\{\alpha_n\}_{n \geq 1}$ *satisfies* $0 \leq \alpha_n < 1$, $\sum_{n=0}^{\infty} \alpha_n = \infty$ *and* $\sum_{n=0}^{\infty} \alpha_n^2 < \infty$.

3.4 Security Applications

In this subsection, we will present some exemplary applications of the minimax-PDS and the WoLF-PDS algorithms in enabling the defender to learn and adapt faster in dynamic security competitions. For this purpose, an exemplary jamming/anti-jamming security game between an EHCS and a jammer (J) is discussed first, followed by a cloud-based security game.

3.4.1 Anti-jamming in Energy Harvesting Communication Systems

Fig. 3.5 The jamming/anti-jamming competition between the EHCS and the jammer

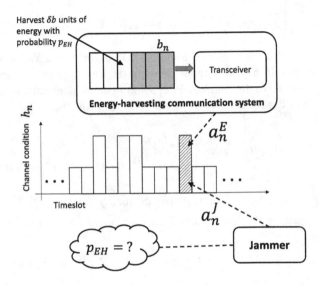

The jamming/anti-jamming competition between the EHCS and the jammer is depicted in Fig. 3.5. The EHCS uses the harvested energy to power its transceiver and at each timeslot, it is assumed that δb units of energy will be harvested with probability p_{EH}. This energy harvesting rate p_{EH} is usually only available to the EHCS as *extra information* and is unknown to the jammer. In the following, we will see how the defender (i.e., the EHCS) can exploit this extra information so as to always maintain a leading edge in anti-jamming.

Due to the dynamics in the wireless channel condition and the variations of the battery status of the EHCS, it is natural to model the jamming/anti-jamming competition as a SG. The state of the game can be defined as $s_n = (b_n, h_n)$, where b_n is the energy level of the EHCS and h_n is the channel power gain at timeslot n. In the rest of the discussion, assume that the state s_n is always observable to both the EHCS and the jammer. The action of the EHCS is denoted by a_n^E, capturing the transmit power of the EHCS at timeslot n. Apparently, the maximum possible transmit power at a certain timeslot is upper bounded by the current energy level b_n of the EHCS. To simply the discussion and focus on the essence of the problem here, discrete energy level is considered here. Particularly, the set of all possible energy level is denoted by $B \triangleq \{b^{(1)}, \ldots, b^{max}\}$ where $b^{(1)} = 0$ and b^{max} equals the battery capacity. After each transmission, the new energy level of the battery is given by

$$b_{n+1} = \begin{cases} \min\left\{b_n - a_n^E + \delta b, b^{max}\right\}, & \text{with probability } p_{EH}, \\ b_n - a_n^E, & \text{otherwise,} \end{cases} \tag{3.17}$$

where it is assumed that δb amount of new energy will be harvested by the EHCS per timeslot with probability p_{EH}. The jammer considered here is assumed to be intelligent and can smartly adjust its jamming power $a_n^J \in A^J$ at each timeslot, where A^J is the action set of the jammer. In addition, the per unit jamming power cost is assumed to be $c_J > 0$. Moreover, for simplicity, we only considered the competition over a single channel, though the method introduced here is general and can be straightforwardly extended to multi-channel applications. The set of the possible power gains of the wireless channel is denoted by $H \triangleq \{h^{(1)}, \ldots, h^{(m)}\}$, and the instant channel power gain $h_n \in H$ at each timeslot n often follows a Markov process. Particularly, denote the probability of the channel power gain switching from $h^{(i)}$ to $h^{(j)}$ by $p(h_{n+1} = h^{(j)} | h_n = h^{(i)}) = p_H(j|i)$. However, in many practical anti-jamming problem, the transition probability $p_H(j|i)$ of the channel power gain is often unknown, leading to an unknown dynamic security game. The goal of the EHCS includes two parts. First, the EHCS aims to transmit as much data as possible. In addition, the EHCS would hope to increase the cost of jamming so as to indirectly deter the jammer. To capture this consideration, the total reward at each timeslot n will be given by

$$r_n = \frac{h_n \cdot a_n^E}{a_n^J + N} + c^J \cdot a_n^J, \tag{3.18}$$

where the first term reflects the reward in terms of throughput (with N the channel noise power) and the second term is the jamming cost to the jammer. Moreover, by taking a zero-sum assumption, the reward of the jammer will be given by

$$r_n^J = -r_n. \tag{3.19}$$

With the problem specified, we will demonstrate how the minimax-PDS and the WoLF-PDS algorithm can be applied by the EHCS to exploit its information advantage so as to achieve the best possible performance. It is worth mentioning that the jammer cannot employ the PDS-learning algorithms due to the lack of knowledge about p_{EH}. To this anti-jamming problem, the PDS can be defined as $\tilde{s}_n = \{\tilde{b}_n, h_n\}$, where

$$\tilde{b}_n \triangleq \min\{b_n - a_n^E + \delta b, b^{max}\}, \tag{3.20}$$

and correspondingly $p^k = p_{EH}$, $p^u = p_H$, $r_n^k = \frac{h_n \cdot a_n^E}{a_n^J + N} + c^J \cdot a_n^J$, and $r_n^u = 0$. With these quantities defined, the minimax-PDS and the WoLF-PDS algorithms depicted in Fig. 3.4 can be readily applied.

In the following, some numerical results will be shown to illustrate the performance gain obtained by using the PDS-learning introduced above. To quantify the performance gain, the following relative performance gain will be used

$$\eta(n) \triangleq \frac{\tilde{r}_{PDS}(n) - \tilde{r}(n)}{\bar{\tilde{r}}(n)} \times 100\%, \tag{3.21}$$

where

$$\tilde{r}_{PDS}(n) \triangleq \frac{1}{n} \sum_{i=1}^{n} r(s_i, a_i, o_i), \tag{3.22}$$

is the average accumulative reward of the PDS-learning algorithm, and $\tilde{r}(n)$ is the average accumulative reward of the conventional algorithm; the denominator $\bar{\tilde{r}}(n)$ denotes the average of $\tilde{r}(n)$ over all Monte Carlo runs. In addition, to quantify the gain in learning speed of the PDS algorithms, the distance between the learned and the optimal quality functions Q_n and Q_*

$$\Delta Q_n \triangleq \frac{||\mathbf{vec}(Q_n - Q_*)||_1}{||\mathbf{vec}(Q_*)||_1} \times 100\%, \tag{3.23}$$

will be used as the performance metric; here $\mathbf{vec}(\cdot)$ and $||\cdot||_1$ denote the vectorization operator and the 1-norm, respectively.

Since the extra information is not available to the jammer, it is assumed that the jammer adopts the conventional minimax-Q algorithm to find its jamming strategy.[2] The learning speed of the minimax-PDS and the conventional minimax-Q algorithms are compared in Fig. 3.6. It can be seen that the learned Q-function of the minimax-PDS algorithm is getting closer to the optimal Q-function much faster than that of the conventional minimax-Q algorithm. For example, after 200 timeslots, the relative distance of the Q-function in the minimax-PDS algorithm reduces to around 15% whereas the relative distance of the conventional minimax-Q algorithm is still above 60%. This evidences that the minimax-PDS algorithm can effectively take advantage of the extra information and provide a significantly faster adaptation to the EHCS. This faster learning capability further leads to enhanced security performance as shown in Fig. 3.7. For example, when $n = 200$, the relative performance gain of the minimax-PDS algorithm over the conventional minimax-Q algorithm is around 30%. Additionally, such relative performance gain will remain above 30% over a long period of time (till $n = 1000$). Eventually, since both algorithms will converge to the optimal Q-function, the relative performance

---

[2]Note that similar performance gain can be observed when the jammer adopts other multi-agent reinforcement algorithms but the corresponding results are omitted here.

gain will vanish. Nonetheless, if the statistics of the dynamic wireless environment
keeps on evolving, such PDS learning algorithm will enable the defender to always
maintain a leading edge in the dynamic security competition. Similar observations
can be made from Fig. 3.8, in which the EHCS adopts the WoLF-PDS algorithm.
Although there is not convergence guarantee for the WoLF-PDS algorithm, usually
it can deliver a much significant performance gain even over the minimax-PDS
algorithm. For example, as shown in Fig. 3.8, the relative performance gain is over
80% when $n = 100$.

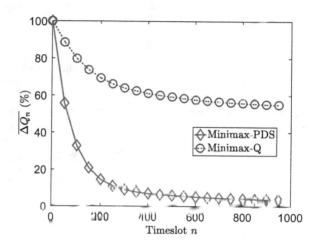

Fig. 3.6 Comparison of the average convergence performance $\overline{\Delta Q_n}$

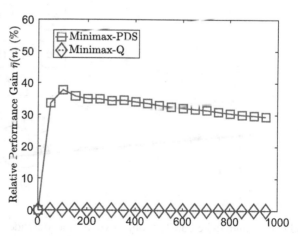

Fig. 3.7 Comparison of the average anti-jamming performance gain $\bar{\eta}(n)$

3.4.2 Cloud-Based Security Game

The second example is concerned with a cloud-based security game. As illustrated
in Fig. 3.9, the defender and the attacker of this security game act simultaneously on
a target network consisting K nodes; each node can be in one of the two states:

Fig. 3.8 WoLF-PDS vs.
WoLF in scenario I

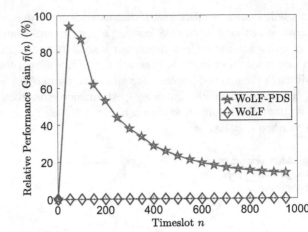

Fig. 3.9 The cloud-based
security game

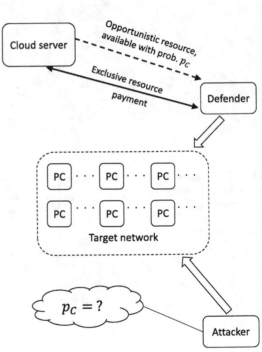

healthy and infected. The objective of the defender is to keep as many nodes
healthy as possible with the least expenditure. It is assumed that the defender
itself does not have the required computational resource to fulfill the desired
defense and hence has to resort to a cloud server for help. Usually, the cloud
server can offer two types of computing resource [20]: The exclusive resource is
guaranteed available at each timeslot but payment is required; additionally, the free
opportunistic computing resource may also be made available to the defender with

a certain probability, depending on the scheduling policy of the cloud server. More specifically, it is assumed that the set of possible amount of opportunistic resource is $\mathscr{C} = \{0, \ldots, c_{max}\}$ with c_{max} the maximum possible amount of opportunistic resource, and that the available opportunistic resource at different timeslot follows a Markov process with transition probability p_C. In practice, it is possible for the cloud server to inform the defender about such statistics, whereas the attacker may not have this information. Again, this scenario leads to a dynamic security game with extra information. In each timeslot of this security game, the defender selects a_n nodes (i.e., the defender's action) in the target network to enforce their security and it is assumed that one unit of computational resource is needed for each selected node. The per unit price of the exclusive resource is denoted by φ_a. On the other hand, the attacker will choose o_n nodes as its targets for malware injection, and the attacking cost of each target node is denoted by φ_o. In addition, a two-phase state dynamics will take place in the considered cloud-based security game. In phase-one, for a certain network node, if it is enforced by the defender but not attacked by the attacker, it will become (or stay in) the healthy state; while if it is attacked by the attacker but not protected by the defender, it will become (or stay in) the infected state. When the defender and the attacker act on the same node, the corresponding state transition probability is denoted by p_{ij}, where $i, j \in \{0, 1\}$ and 0 and 1 correspond to the healthy and the infected states, respectively. If neither the defender nor the attacker acts on a certain node, then its state will stay unchanged. The second phase aims to capture the impact of malware spreading. Specifically, in this spreading phase, any node in a healthy state may be infected by an infected node (if there is any) with a certain probability, denoted by p_{inf}. According to the amount of available opportunistic resource and the status of the network nodes, the state of this security game can be defined as $s_n = (c_n, \mathbf{1}_1, \ldots, \mathbf{1}_K)$ where $\mathbf{1}_i \in \{0, 1\}$ reflects whether or not the ith node is healthy. By assuming that there are in total K nodes in the network and denoting by k_n the number of infected nodes, the reward of the defender in this security game is modeled by

$$r_n = (K - k_n) - \varphi_a \cdot \max\{a_n - \iota_n, 0\} + \varphi_o \cdot o_n, \qquad (3.24)$$

where the first term corresponds to the reward of keeping nodes healthy; the second term captures the cost of purchasing exclusive computational resource from the cloud server; additionally, under a zero-sum assumption, the last term reflects the cost of the attacker. The reward of the attacker is simply given by

$$r_n^A = -r_n. \qquad (3.25)$$

To apply the minimax-PDS and the WoLF-PDS algorithms, one can define a PDS $\tilde{s}_n = (\mathbf{1}_1, \ldots, \mathbf{1}_K, \tilde{c}_n)$ where $\tilde{c}_n \triangleq c_{n+1}$. Additionally, let the known part of the state transition probability be $p^k = p_C$ and the unknown part p^u is determined by the node state transition probabilities p_{ij} for $i, j = 0, 1$ and the malware spreading probability p_{inf}. Accordingly, the known part of the reward is set to be

$r_n^k = r_n$ (defined in (3.25)) and the unknown part of the reward in the PDS-learning algorithms is set to zero. The corresponding performance are shown in Figs. 3.10 and 3.11 for the minimax-PDS and the WoLF-PDS algorithms, respectively. Particularly, as shown in Fig. 3.10, when the attacker adopts the conventional minimax-Q algorithm to find its strategy, the minimax-PDS algorithm enables the defender to gain substantial extra reward as compared to the conventional minimax-Q algorithm. For example, at $n = 1000$, the relative performance gain almost reaches 80%. Similar observations can be made from Fig. 3.11 in which the attacker is assumed to adopt the conventional WoLF algorithm, and again, PDS-learning allows the defender to fully leverage its information advantage and achieve a significantly faster learning of the optimal defense strategy.

Fig. 3.10 Minimax-PDS vs. minimax-Q in cloud-based security game (with a minimax-Q attacker)

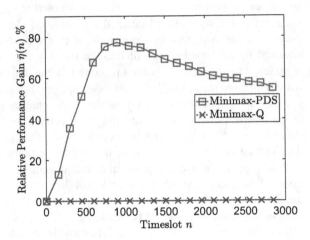

Fig. 3.11 WoLF-PDS vs. WoLF in cloud-based security game (with a WoLF attacker)

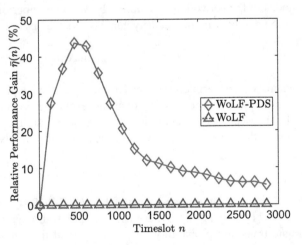

3.5 Summary

In this chapter, dynamic security games with extra information have been investigated. To handle the extra information, the PDS-learning technique has been incorporated into conventional multi-agent RL algorithms, leading to two new algorithms, minimax-PDS and WoLF-PDS. These two algorithms can enable the defender to fully exploit its extra information and achieve faster learning and adaptation in dynamic security games. In addition, the EHCS anti-jamming problem and the cloud-based security game are taken as two examples to illustrate the effectiveness of these two algorithms. The corresponding numerical results show that, by using these two new algorithms, the defender's security performance can be considerably improved.

References

1. N. Mastronarde, M. van der Schaar, Fast reinforcement learning for energy-efficient wireless communication. IEEE Trans. Signal Process. **59**(12), 6262–6266 (2011)
2. S. Sudevalayam, P. Kulkarni, Energy harvesting sensor nodes: survey and implications. IEEE Commun. Surv. Tutorials **13**(3), 443–461 (2011)
3. R. Prasad, S. Devaschapathy, V. Rao, J. Vazifehdan, Reinformation in the ambiance: devices and networks with energy harvesting. IEEE Commun. Surv. Tutorials **16**(1), 195–213 (2014)
4. A. Mpitziopoulos, D. Gavalas, C. Konstantopoulos, G. Pantziou, A survey on jamming attacks and countermeasures in WSNs. IEEE Commun. Surv. Tutorials **11**(4), 42–56 (2009)
5. M. Strasser, S. Capkun, C. Popper, M. Cagalj, Jamming-resistant key establishment using uncoordinated frequency hopping, in *Proceedings of IEEE SP (Oakland)* (2008)
6. C. Li, H. Dai, L. Xiao, P. Ning, Communication efficiency of anti-jamming broadcast in large-scale multi-channel wireless networks. IEEE Trans. Signal Process. **60**(10), 5281–5292 (2012)
7. W. Xu, W. Trappe, Y. Zhang, Anti-jamming timing channels for wireless networks, in *Proceedings of ACM WiSec* (2008)
8. Y. Liu, P. Ning, BitTrickle: defending against broadband and high-power reactive jamming attacks, in *Proceedings of IEEE INFOCOM* (2012)
9. P. Tague, Improving anti-jamming capability and increasing jamming impact with mobility control, in *Proceedings of IEEE MASS* (2010)
10. X. He, H. Dai, P. Ning, Dynamic adaptive anti jamming via controlled mobility, in *Proceedings of IEEE CNS*, National Harbor, MD (2013)
11. A. György, T. Linder, G. Lugosi, G. Ottucsák, The on-line shortest path problem under partial monitoring. J. Mach. Learn. Res. **8**, 2369–2403 (2007)
12. Q. Wang, P. Xu, K. Ren, X.Y. Li, Towards optimal adaptive UFH-based anti-jamming wireless communication. IEEE J. Sel. Areas Commun. **30**(1), 16–30 (2012)
13. H. Li, Z. Han, Dogfight in spectrum: combating primary user emulation attacks in cognitive radio systems, Part I: known channel statistics. IEEE Trans. Wirel. Commun. **9**(11), 3566–3577 (2010)
14. M. Armbrust, A. Fox, R. Griffith, A.D. Joseph, R. Katz, A. Konwinski, G. Lee, D. Patterson, A. Rabkin, I. Stoica, A view of cloud computing. Commun. ACM **53**(4), 50–58 (2010)
15. J. Oberheide, K. Veeraraghavan, E. Cooke, J. Flinn, F. Jahanian, Virtualized in-cloud security services for mobile devices, in *ACM Workshop on Virtualization in Mobile Computing*, Breckenridge, CO, June 2008

16. V. Varadharajan, U. Tupakula, Security as a service model for cloud environment. IEEE Trans. Netw. Serv. Manag. **11**(1), 60–75 (2014)
17. W. Yassin, N.I. Udzir, Z. Muda, A. Abdullah, M.T. Abdullah, A cloud-based intrusion detection service framework, in *Proceedings of IEEE CyberSec*, Kuala Lumpur, June 2012
18. Y. Meng, W. Li, L. Kwok, Design of cloud-based parallel exclusive signature matching model in intrusion detection, in *Proceedings of IEEE PHCC and EUC*, Hunan, Nov 2013
19. X. He, H. Dai, P. Ning, Faster learning and adaptation in security games by exploiting information asymmetry. IEEE Trans. Signal Process. **64**(13), 3429–3443 (2016)
20. T. He, S. Chen, H. Kim, L. Tong, K. Lee, Scheduling parallel tasks onto opportunistically available cloud resources, in *Proceedings of IEEE CLOUD*, Honolulu, HI, June 2012

Chapter 4
Dynamic Security Games with Incomplete Information

4.1 Introduction

In this chapter, we continue our exploration of dynamic security games with information asymmetry. As compared to the previous chapter, the discussions in this chapter focus on the complementary scenarios where the defender lacks information about the ongoing security competitions. Such scenarios exist in many practical security problems. For example, an IDS is usually unaware of the purpose of the potential intruder as well as its most likely attacks. As another example, in an adversarial network environment, a legitimate transmitter may not have information about whether or not other network users are malicious.

The framework of Bayesian SG [1] will be employed in this chapter to model and analyze such incomplete information dynamic security problems. Accordingly, a new algorithm, termed Bayesian Nash-Q, that allows the defender to infer the missing information based on repeated interactions with the attacker and the dynamic environment will be presented. This algorithm is a natural combination of the conventional repeated Bayesian games and the Nash Q algorithm introduced in Chap. 1. For this reason, our discussion starts from reviewing some elementary concepts of Bayesian games, and then the Bayesian SG model and the Bayesian Nash-Q algorithm will be discussed in details.

Two concrete examples will be taken to illustrate the application of the Bayesian SG model and the corresponding Bayesian Nash-Q algorithm. The first example is concerned with the dynamic IDS configuration problem [2, 3]. Particularly, as different information systems are permeating through our daily life, deploying reliable IDSs are becoming increasingly important. However, this is a non-trivial mission as practical IDSs are only equipped limited resource and hence cannot load all the intrusion detection algorithms simultaneously, whereas the objective and intended attack of the intruder are often unknown beforehand. To address this problem, the Bayesian Nash-Q algorithm can be employed to identify the intruder's objective and adequately configure the IDS accordingly so as to achieve

© The Author(s) 2018
X. He, H. Dai, *Dynamic Games for Network Security*, SpringerBriefs in Electrical and Computer Engineering, https://doi.org/10.1007/978-3-319-75871-8_4

the most efficient defense. The second example is about spectrum access in adversarial wireless communication networks. As the wireless networks becoming more accessible to the open public, malicious wireless users may easily enter the networks to disrupt the normal data transmission of legitimate users [4–9]. Consequently, it is important for the legitimate users to identify whether or not a certain peer is malicious so as to adjust their spectrum access behaviors accordingly. Apparently, this conforms to the setting assumed by the Bayesian SG framework and the Bayesian Nash-Q algorithm can be applied to enable legitimate wireless users to achieve the desired objective.

4.2 Background on Repeated Bayesian Game

In this section, some background on Bayesian games and repeated Bayesian games [10, 11] is introduced to smooth the later discussions.

Bayesian games are usually concerned with games in which the *types* of the players are unknown to the public, leading to games with *incomplete information*. Particularly, the type of a player specifies the purpose of the player and is uniquely determined by the player's reward function. In the sequel of this chapter, we will focus on a special class of two-player Bayesian games where only one player's type is private, which corresponds to the scenario of unknown attacker type in practical security problems; interested readers may refer to, for example [10, 11], for the discussions of more general Bayesian games. The Bayesian games considered here are different from the conventional two-player games discussed in Chap. 1 in the following aspects:

- The type θ of one player is unknown to the public and is assumed to belong to a finite discrete set $\Theta = \{\theta^{(1)}, \theta^{(2)}, \ldots, \theta^{(|\Theta|)}\}$. This player will be called, the *informed player* and the other player will be called, the *uninformed player* in the rest of this chapter;
- The reward functions of both players may depend on not only their actions but also θ;
- The uninformed player has a *belief* about its opponent's type θ, and the belief is represented by a probability distribution over Θ.

When a Bayesian game is played repeatedly, the uninformed player (e.g., the defender in security applications) can gradually learn the type of its opponent (e.g., the attacker) by leveraging Bayesian learning, as illustrated in Fig. 4.1. The basic idea is as follows. The uninformed player can first compute the equilibrium strategies for all possible types of opponents as in conventional two-player games. In the following discussions, the equilibrium strategy of type-γ opponent is denoted by $\tilde{\pi}^\gamma$ and the probability of a type-γ opponent taking a certain action o_n is denoted by $\tilde{\pi}^\gamma(o_n)$. In addition, the uninformed player maintains a belief b_n^γ, indicating the posterior probability that the opponent is of type-γ given the observation up to timeslot n. At each timeslot, after observing the opponent's action o_n, the uninformed player can update its belief b_{n+1}^γ by using the Bayes' theorem as follows

$$b_{n+1}^{\gamma} = \frac{b_n^{\gamma} \cdot \tilde{\pi}^{\gamma}(o_n)}{\sum_{\gamma'=1}^{|\Theta|} b_n^{\gamma'} \cdot \tilde{\pi}^{\gamma'}(o_n)}. \tag{4.1}$$

In infinitely repeated Bayesian games, it can be shown that the maintained belief b_{n+1}^{γ} will eventually converge to the true value [9, 12].

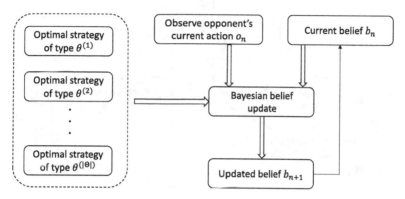

Fig. 4.1 Belief update in repeated Bayesian games

4.3 Bayesian SGs

The Bayesian SG is a generalization of the repeated Bayesian game introduced above by incorporating the notion of state. In this section, the Bayesian SG model will be introduced, along with the Bayesian Nash-Q algorithm that can guide the uninformed player to derive its equilibrium strategy in incomplete information SGs.

4.3.1 Bayesian SG Model

The Bayesian SG model depicted in Fig. 4.2 is quite similar to the conventional SG presented in Chap. 1. The main difference is the additional belief update process (c.f. the grey box in Fig. 4.2), which takes the observed current state and the opponent's action as input. In the Bayesian SG model, if the opponent is rational, its action should be aligned to its purpose and hence may disclose some clue about its type. Consequently, to identify the opponent's type, the defender can construct a *belief vector* $\mathbf{b} = [b^{\gamma}]_{\gamma \in \Gamma}$, where Γ represents the set of all possible opponent types and b^{γ} represents the defender's belief that the opponent is of type-γ and admits

$$\sum_{\gamma \in \Gamma} b^{\gamma} = 1. \tag{4.2}$$

To make the problem tractable, we further make an assumption that the number of possible opponent types $|\Gamma|$ is finite.

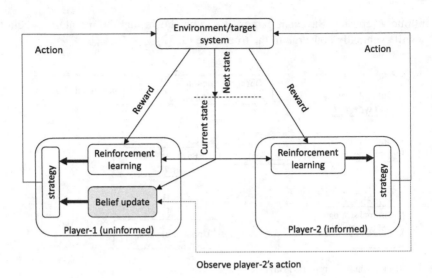

Fig. 4.2 Bayesian SG

In addition, it is assumed in the Bayesian SG that the type-dependent reward functions may are known to the public. As a result, although the uninformed player is not aware of the instant reward of the unknown type opponent, it can derive the instant reward of any given type of opponent. Considering this, in addition to the belief vector, the uninformed player also maintains sets of quality and value functions $\{Q^{D,\gamma}, \tilde{Q}^{\gamma}\}_{\gamma \in \Gamma}$ and $\{V^{D,\gamma}, \tilde{V}^{\gamma}\}_{\gamma \in \Gamma}$, respectively. Particularly, for each $\gamma \in \Gamma$, the optimal $Q^{D,\gamma}$ is defined as

$$Q_*^{D,\gamma}(s, l, a) \triangleq R^D(s, l, a) + \mathbb{E}_{s'}\left[\beta \cdot V_*^{D,\gamma}(s')\right], \qquad (4.3)$$

where s, l, and a denote the state, the uninformed player's action and that of the opponent. The type-dependent value function $V_*^{D,\gamma}$ is defined as

$$V_*^{D,\gamma}(s) \triangleq \text{NASH}^D\left(Q_*^{D,\gamma}(s, \cdot, \cdot), \tilde{Q}_*^{\gamma}(s, \cdot, \cdot)\right), \qquad (4.4)$$

where the operator NASH^D takes two matrices $Q_*^{D,\gamma}(s, \cdot, \cdot)$ and $\tilde{Q}_*^{\gamma}(s, \cdot, \cdot)$ as the input and outputs the value of the game at Nash equilibrium for the uninformed player. Similarly, the optimal $\tilde{Q}^{D,\gamma}$ and $\tilde{V}^{D,\gamma}$ are defined respectively as

$$\tilde{Q}_*^{\gamma}(s, l, a) \triangleq R^{\gamma}(s, l, a) + \mathbb{E}_{s'}\left[\beta \cdot \tilde{V}_*^{\gamma}(s')\right], \qquad (4.5)$$

and

$$\mathring{V}_*^\gamma(s) \triangleq \text{NASH}^\gamma \left(Q_*^{D,\gamma}(s, \cdot, \cdot), \tilde{Q}_*^\gamma(s, \cdot, \cdot) \right). \tag{4.6}$$

Based on the quantities defined above, we will demonstrate how to derive the optimal strategy in Bayesian SG in the next subsection.

4.3.2 Bayesian Nash Q-Learning

In this subsection, the Bayesian Nash Q-learning algorithm that enables the uninformed player to gradually learn its optimal strategy in possibly unknown dynamics environments will be studied.

Specifically, the uninformed player can adopt the following reinforcement procedure to learn the Q- and value functions defined in (4.3)–(4.6)

$$Q_{n+1}^{D,\gamma}(s, l, a) = \begin{cases} (1 - \alpha_n) Q_n^{D,\gamma}(s, l, a) + \alpha_n \left[r_n^D + \beta \cdot V_n^{D,\gamma}(s_{n+1}) \right], \\ \qquad\qquad\qquad \text{for } (s, l, a) = (s_n, l_n, a_n), \\ Q_n^{D,\gamma}(s, l, a), \qquad\qquad \text{otherwise}, \end{cases} \tag{4.7}$$

$$V_{n+1}^{D,\gamma}(s) = \text{NASH}^D \left(Q_{n+1}^{D,\gamma}(s, \cdot, \cdot), \tilde{Q}_{n+1}^\gamma(s, \cdot, \cdot) \right), \tag{4.8}$$

$$\tilde{Q}_{n+1}^\gamma(s, l, a) = \begin{cases} (1 - \alpha_n) \tilde{Q}_n^\gamma(s, l, a) + \alpha_n \left[R^\gamma(s, l, a) + \beta \cdot \tilde{V}_n^\gamma(s_{n+1}) \right], \\ \qquad\qquad\qquad \text{for } (s, l, a) = (s_n, l_n, a_n), \\ \tilde{Q}_n^\gamma(s, l, a), \qquad\qquad \text{otherwise}, \end{cases} \tag{4.9}$$

and

$$\tilde{V}_{n+1}^\gamma(s) = \text{NASH}^\gamma \left(Q_{n+1}^{D,\gamma}(s, \cdot, \cdot), \tilde{Q}_{n+1}^\gamma(s, \cdot, \cdot) \right). \tag{4.10}$$

Note that (4.9) and (4.10) are conducted by the uninformed player to imitate the learning procedure of the opponent; this imitation procedure requires only local information available to the uninformed player. With the above quantities, the uninformed player uses $\tilde{Q}^\gamma$ and $\tilde{V}^\gamma$ as the virtual quality and value functions to infer the most likely strategy $\tilde{\pi}^\gamma$ from a type-γ opponent by

$$\tilde{\pi}_{n+1}^\gamma(s, \cdot) = \arg \text{NASH}^\gamma \left(Q_{n+1}^{D,\gamma}(s, \cdot, \cdot), \tilde{Q}_{n+1}^\gamma(s, \cdot, \cdot) \right), \tag{4.11}$$

where the operator NASH^γ holds similar definition as NASH^D but generates the value of the game for the opponent. Once $\tilde{\pi}^\gamma$'s are obtained for all $\gamma \in \Gamma$, the uninformed player compares these strategies and the observed action from the opponent and then updates the belief vector using Bayes' formula as follows:

$$b_{n+1}^{\gamma} = \frac{b_n^{\gamma} \cdot f_n^{\gamma}}{\sum_{\gamma' \in \Gamma} b_n^{\gamma'} \cdot f_n^{\gamma'}}, \quad \forall \gamma \in \Gamma, \tag{4.12}$$

where $f_n^{\gamma'}$ is the likelihood that a type-γ' intruder takes action a_n at the current state s_n and is given by

$$f_n^{\gamma'} = (1 - p_{explr}) \cdot \tilde{\pi}_n^{\gamma'}(s_n, a_n) + \frac{p_{explr}}{N}, \quad \forall \gamma' \in \Gamma. \tag{4.13}$$

In the above equation, p_{explr} is the exploration probability and N is the number of possible actions of the uninformed player. In addition, the uninformed player derives the best response against each possible type of opponent based on $Q^{D,\gamma}$ and $V^{D,\gamma}$ as follows

$$\pi_{n+1}^{IDS,\gamma}(s, \cdot) = \arg \mathrm{NASH}^D \left(Q_{n+1}^{D,\gamma}(s, \cdot, \cdot), \tilde{Q}_{n+1}^{\gamma}(s, \cdot, \cdot) \right). \tag{4.14}$$

Further based on the current belief vector $\mathbf{b}_n = [b_n^{\gamma}]_{\gamma \in \Gamma}$, a weighted strategy is constructed by the uninformed player as follows

$$\pi_n^D = \sum_{\gamma \in \Gamma} b_n^{\gamma} \cdot \pi_n^{D,\gamma}. \tag{4.15}$$

The above procedure of finding the optimal strategy in incomplete information SG is known as the Bayesian Nash-Q algorithm [3] and is summarized in Fig. 4.3. It can be seen that the Bayesian Nash-Q algorithm is a generalization of the conventional Nash-Q algorithm reviewed in Chap. 1 and hence, the convergence is not guaranteed for the Bayesian Nash-Q algorithm in general. Nonetheless, in some special cases, as will be shown later in Sect. 4.4.2, the Bayesian Nash-Q algorithm will converge. In addition, it is worth mentioning that the learning procedure described in (4.7)–(4.10) for different opponent types can be executed in parallel, and consequently, learning efficiencies of the Bayesian Nash-Q algorithm and the conventional Nash-Q algorithm are comparable when the uninformed player is equipped parallel computing devices.

4.4 Security Applications

In this section, two exemplary applications of the above Bayesian SG framework will be illustrated. The first example concerns the dynamic configuration problem of IDS and the second one targets the spectrum access issue in adversarial wireless networks.

4.4.1 Dynamic Intrusion Detection System Configuration

The objective of this subsection is to demonstrate the application of the Bayesian Nash-Q algorithm in addressing the dynamic IDS configuration problem. Readers who are not familiar with IDS may refer to, for example, [13, 14] and the references therein for background knowledge.

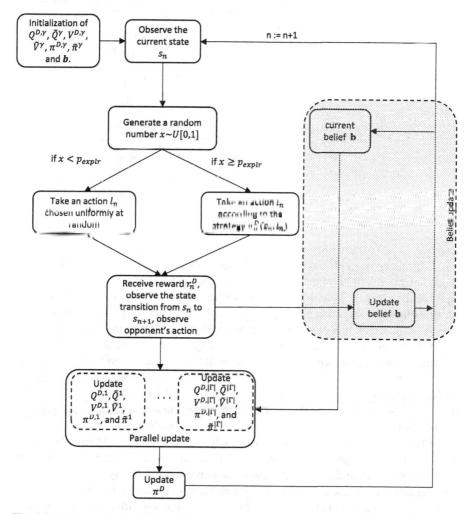

Fig. 4.3 The Bayesian Nash-Q algorithm

Although a pool of abundant intrusion detection algorithms have developed in the literature [13–19], none of them is a panacea in that each detection algorithm is only good at detecting a certain type of intrusions. On the other hand, it often takes a non-negligible amount of computational and storage resource to execute these intrusion detection algorithms. For these reasons, IDS configuration, which concerns about

Fig. 4.4 Dynamic intrusion
detection system
configuration

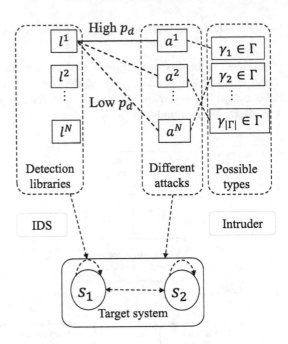

optimally selecting suitable detection algorithms with limited available security
resource, is of paramount importance in practical applications. As illustrated in
Fig 4.4, this is a highly non-trivial task, since (1) the state of the target system
protected by the IDS often changes over time and (2) the intruder's type (or
equivalently, purpose) can rarely be known beforehand. In the following, we will
illustrate how the Bayesian Nash-Q algorithm discussed in the previous section can
be applied to crack this challenging problem. To this end, the interaction between
the IDS, the unknown type intruder, and the target system will be modeled as an
SG with incomplete information, in which the IDS is the uninformed player and the
intruder is the informed one.

Some notations and assumptions are clarified first to smooth the discussions
thereafter. Specifically, as shown in Fig 4.4, the IDS is assumed to equipped
with N libraries of intrusion detection algorithms, denoted by $\mathscr{L} = \{l^1, \ldots, l^N\}$,
while there are N different possible attacks from the adversary, denoted by $\mathscr{A} =
\{a^1, \ldots, a^N\}$. Without loss of generality, it is assumed that the IDS can only load
one library at a time due limited security resource, and similarly, the attacker can
only launch one attack each time. Additionally, it is assumed that there is a one-
to-one correspondence between the intrusion detection algorithm libraries and the
potential attacks. More specifically, it is assumed that library l^i can detect attack a^i
with a high probability $p_{i,i}$ but other attacks $\{a^j\}$ $(j \neq i)$ with low probabilities
$\{p_{i,j}\}$, which reflects the fact that none of the existing intrusion detection algorithm
is perfect. To make the problem tractable, it is also assumed that the set Γ of possible
intruder types is of finite cardinality. Depending on the actions taken by the IDS and
the attacker, the state of the target system will change over time following a certain
probability distribution. For example, when the library loaded by the IDS matches

with the attack, the target system is likely to remain at (or change to) the healthy state. From the IDS's viewpoint, it aims to load the correct library so as to catch the intruder with a probability as high as possible. With this consideration, we model the reward function of the IDS as follows:

$$R^{IDS}(s, l^i, a^j) = p_{i,j} \cdot w_{s,j}^{IDS}, \qquad (4.16)$$

where $w_{s,j}^{IDS}$ is a weighting factor that captures the importance of detecting attack a^j at state s. For example, when the target system is offline and updating its software, a small weighting factor will be assigned to a network congestion attack [20] due to its negligible influence. On the other hand, depending on its type, different intruders may have different preferences over the N possible attacks. For a type-γ intruder (with $\gamma \in \Gamma$), its reward function can be modeled as

$$R^\gamma(s, l^i, a^j) = (1 - p_{i,j}) \cdot w_{s,j}^\gamma. \qquad (4.17)$$

The rationale behind this modeling is that the first term $(1 - p_{i,j})$ captures the probability of successful attack and the second term denotes the profit to the intruder for successfully launching attack a^j at state s. It is worth mentioning that, although the type of intruder is unknown to the IDS, it is reasonable to assume that the set of payoff functions $\{R^{\gamma}\}_{\gamma \in \Gamma}$ are publicly known

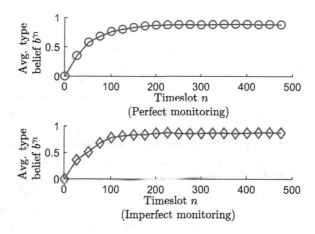

Fig. 4.5 Intruder type identification

With the above modeling, the Bayesian Nash-Q algorithm introduced in the previous section can be readily applied to address the dynamic IDS configuration problem, and the corresponding numerical results are presented below. In particular, it is assumed that there are three possible types of intruders, denoted by γ_1, γ_2 and γ_3, respectively; the true type of the intruder is γ_1. To facilitate performance comparison, the average accumulative reward defined as

$$\bar{r}_n^{IDS} = \frac{1}{n} \sum_{i=1}^{n} r_i^{IDS}, \qquad (4.18)$$

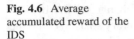

Fig. 4.6 Average accumulated reward of the IDS

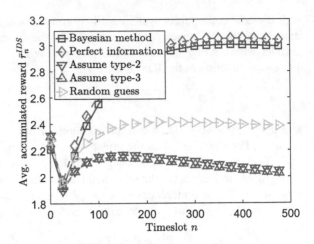

is adopted as the metric. First, let us examine the performance of the Bayesian Nash-Q algorithm in a somewhat ideal case in which the IDS can observe the intruder's action at each timeslot (i.e., the perfect monitoring case); the purpose is to focus on the demonstration of the intruder type identification and performance enhancement capability of the Bayesian Nash-Q algorithm. The type identification performance of the Bayesian Nash-Q algorithm is presented in the first subplot of Fig. 4.5. It can be seen that the IDS's belief that the intruder of type-1 is monotonically increasing and after around $n = 100$ timeslots, the belief b^{γ_1} exceeds 0.8. This indicates that by adopting the Bayesian Nash-Q algorithm, the IDS is able to infer the intruder's type with high fidelity. The fact that the belief b^{γ_1} does not reach 1 even after 400 timeslots deserves some further explanation: The Bayes' update in (4.33) converges only when the estimated strategy by the IDS and the true strategy of the intruder are identical. However, the considered dynamic IDS configuration problem is a general-sum security game (c.f. (4.16) and (4.17)). For such games, neither the conventional Nash-Q algorithm nor the Bayesian Nash-Q algorithm is guaranteed to converge. Consequently, the estimated type-dependent strategy by the IDS and the strategy adopted by the intruder may not be perfectly aligned. Even though not perfect, the type identification achieved by the Bayesian Nash-Q algorithm is still accurate enough to induce a substantial improvement in security performance. This is shown in Fig. 4.6. Particularly, several performance curves are compared in Fig. 4.6. The "Bayesian method" curve corresponds to the security performance of using the Bayesian Nash-Q algorithm. The "Perfect information" curve corresponds to the security performance when the IDS is aware of the intruder's type, which serves as the upper bound of the performance. The "Assume type-2" curve represents the security performance when the IDS incorrectly treats the intruder as type-2 and uses the conventional Nash-Q algorithm to guide its configuration. The "Assume type-3" curve holds similar interpretation for falsely assuming a type-3 intruder. The "Random guess" curve corresponds to the case when the IDS takes a random guess about the intruder's type (i.e., always assumes $b^{\gamma_1} = b^{\gamma_2} = b^{\gamma_3} = \frac{1}{3}$). Several observations from Fig. 4.6 are in order. First, the performance of the Bayesian Nash-Q algorithm substantially exceeds those when the IDS assumes an incorrect intruder type or takes a random guess. Also, due to the accurate type identification, the performance of the Bayesian Nash-Q algorithm is very close to that of the "Perfect information" case; in this sense, the Bayesian Nash-Q algorithm is nearly optimal here.

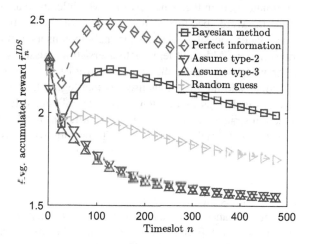

Fig. 4.7 Average accumulated reward of the IDS (with imperfect monitoring)

In the scenario considered above, the IDS is assumed to be able to observe the intruder's action in every timeslot. Now, let us examine a more practical scenario in which the IDS observes the intruder's action if and only if intrusion detection is successful; from a technical viewpoint, this will gives a SG with incomplete information and imperfect monitoring, which is still an open problem. Specifically, when the IDS loads library l^i and the intruder launches the attack a^j, the IDS will observe the intruder's action with probability $p_{i,j}$. In addition, when the IDS fails to detect the intrusion, it makes a random conjecture about the intruder's action. The corresponding type identification and security performances are shown in the second subplot of Figs. 4.5 and 4.7, respectively. It can be seen from the second subplot of Fig. 4.5 that, even with imperfect monitoring, the IDS can still obtain a fairly accurate belief about the intruder's type by using the Bayesian Nash-Q algorithm. As a result, the IDS achieves a substantially better performance that the "Type-2", "Type-3" and "Random guess" cases as shown in Fig. 4.7. But due to imperfect monitoring, a performance gap between the cases of "Bayesian method" and the ideal "Perfect information" can be observed.

4.4.2 Spectrum Access in Adversarial Environment

In this subsection, spectrum access in adversarial environments is taken as another example to illustrate the security application of the Bayesian Nash-Q algorithm.

Spectrum access in adversarial environment concerns the following problem. In a multi-channel wireless communication network, each user is able to select one channel per timeslot for communication. Due to environmental variations, the power gain h^i of each wireless channel-i may change over time, making the spectrum access problem highly non-trivial. Additionally, malicious users may coexist with legitimate users in the same network, which leads to an adversarial spectrum access problem. More specifically, the objective of the legitimate users is to transmit as much data as possible whereas the malicious users aim to jam the legitimate users' transmission. Although legitimate users' transmissions may also collide with each other when they select the same spectrum band, the key difference between the legitimate and the malicious users is that legitimate users intend to avoid collision whereas malicious users intend to create collisions. It is apparent that the channel selection strategies will be completely different for the cases of facing a legitimate peer user and a malicious peer user. To provide some insights about this adversarial spectrum access problem, the basic two-user case is examined here. Particularly, it is assumed that there are two users, denoted by user-I and user-II, and we will analyze the problem from user-I's perspective. In addition to the channel power gain dynamics, another challenge in this problem is that, due to the coexistence of legitimate and malicious users, user-I is usually not aware of user-II's type (i.e., friend or enemy). Fortunately, by modeling the adversarial spectrum access problem as an incomplete information SG, the Bayesian Nash-Q algorithm can enable user-I to infer the possible type of user-II by monitoring the type-driven actions from user-II. By default, we assume that user-I is legitimate. In the discussion hereafter, the following reward function of user-I is assumed:

$$r_n^I = R^I(s_n, a_n^I, a_n^{II}) \triangleq B \cdot \log\{1 + \frac{P \cdot h_n^{a_n^I}}{N}\} \cdot \mathbf{1}_{\{a_n^I \neq a_n^{II}\}}. \tag{4.19}$$

The above modeling intends to capture the throughput of user-I. Specifically, $s_n \triangleq \{h_n^{i_1}, \ldots, h_n^{i_k}\}$ is the state of this SG, defined based on the channel power gains of the k wireless channels; a_n^I and a_n^{II} are the selected channels by user-I and user-II, respectively; B, P and N denote the channel bandwidth, the transmit power, and the noise power, respectively; $\mathbf{1}_{\{E\}}$ is the indicator function, which equals one when the event E is true and zero otherwise.

As to user-II, its reward function apparently depends on its type. When user-II is of the enemy type (i.e., it is a malicious user), its objective is to disrupt user-I's transmission as much as possible rather than maximize its own throughput. Therefore, the more user-I gains, the less user-II obtains. This naturally implies a zero-sum relation between the reward functions of user-I and user-II, and mathematically, it implies

$$R^{II,(e)} = -R^I, \tag{4.20}$$

where the superscript (e) is used to indicate that it is the reward of an enemy type user-II. In this case, it can be verified that the learning procedure (4.7) and (4.8) in the Bayesian Nash-Q algorithm coincides with that in the conventional minimax-Q algorithm [21]. More specifically, for the adversarial spectrum access problem, one can define the quality and the value functions as follows:

$$Q_*^{(e)}(s, a^I, a^{II}) \triangleq R^I(s, a^I, a^{II}) + \mathbb{E}_{s'}\left[\beta \cdot V_*^{(e)}(s')\right], \tag{4.21}$$

and

$$V_*^{(e)}(s) \triangleq \max_{\pi^I(s)} \min_{a^{II} \in \mathscr{A}^{II}} \sum_{a^I} Q_*^{(e)}(s, a^I, a^{II})\pi_{a^I}^I(s). \tag{4.22}$$

Accordingly, reinforcement learning can be conducted by following the minimax-Q algorithm as follows:

$$Q_{n+1}^{(e)}(s, a^I, a^{II}) = \begin{cases} (1 - \alpha_n)Q_n^{(e)}(s, a^I, a^{II}) + \alpha_n\left[r_n^I + \beta \cdot V_n^{(e)}(s_{n+1})\right] \\ \qquad\qquad , \text{ for } (s, a^I, a^{II}) = (s_n, a_n^I, a_n^{II}), \tag{4.23} \\ Q_n^{(e)}(s, a^I, a^{II}), \qquad\qquad \text{otherwise}, \end{cases}$$

$$V_{n+1}^{(e)}(s) = \max_{\pi^I(s)} \min_{a^{II} \in \mathscr{A}^{II}} \sum_{a^I} Q_{n+1}^{(e)}(s, a^I, a^{II})\pi_{a^I}^I(s). \tag{4.24}$$

$$\pi_{n+1}^{(e)}(s) = \arg\max_{\pi^I(s)} \min_{a^{II} \in \mathscr{A}^{II}} \sum_{a^I} Q_{n+1}^{(e)}(s, a^I, a^{II})\pi_{a^I}^I(s). \tag{4.25}$$

Accordingly, the strategy of user-I is updated by

$$\pi_{n+1}^{(e)}(s) = \arg\max_{\pi^I(s)} \min_{a^{II} \in \mathscr{A}^{II}} \sum_{a^I} Q_{n+1}^{(e)}(s, a^I, a^{II})\pi_{a^I}^I(s). \tag{4.26}$$

When user-II is of the friend type, its objective is the same as that of user-I and, therefore, its reward function is given by

$$r_n^{II,(f)} = R^{II,(f)}(s_n, a_n^I, a_n^{II}) \triangleq B \cdot \log\{1 + \frac{P \cdot h_n^{a_n^{II}}}{N}\} \cdot \mathbf{1}_{\{a_n^{II} \neq a_n^I\}}. \tag{4.27}$$

Note that, when both users are legitimate, they may be willing to cooperate with each other to achieve a better social welfare. That being said, user-I aims to optimize a long-term reward $\mathbb{E}\left[\sum\limits_{n=1}^{\infty} \beta^n \cdot \left(r_n^I + r_n^{II,(f)}\right)\right]$. In such case, it can be verified that the learning procedure (4.7) and (4.8) in the Bayesian Nash-Q algorithm is equivalent to

$$Q_{n+1}^{(f)}(s, a^I, a^{II}) = \begin{cases} (1 - \alpha_n) Q_n^{(f)}(s, a^I, a^{II}) + \alpha_n \left[(r_n^I + R^{II,(f)}(s, a^I, a^{II})) \right. \\ \left. + \beta \cdot V_n^{(f)}(s_{n+1}) \right], \text{ for } (s, a^I, a^{II}) = (s_n, a_n^I, a_n^{II}), \\ Q_n^{(f)}(s, a^I, a^{II}), \qquad\qquad\qquad\qquad\qquad \text{otherwise,} \end{cases}$$

$$V_{n+1}^{(f)}(s) = \max_{\pi^I(s)} \max_{a^{II} \in \mathscr{A}^{II}} \sum_{a^I} Q_{n+1}^{(f)}(s, a^I, a^{II}) \pi_{a^I}^I(s). \qquad (4.28)$$

Accordingly, the strategy of user-I is updated by

$$\pi_{n+1}^{(f)}(s) = \arg\max_{\pi^I(s)} \max_{a^{II} \in \mathscr{A}^{II}} \sum_{a^I} Q_{n+1}^{(f)}(s, a^I, a^{II}) \pi_{a^I}^I(s). \qquad (4.29)$$

In addition to the above, user-I has to conduct type identification dictated by the Bayesian Nash-Q algorithm. Specifically, user-I maintains a two-dimension belief vector $\mathbf{b} = [b^{(f)}, b^{(e)}]$ in which $b^{(f)}$ and $b^{(e)}$ represent its beliefs that user-II is enemy type and friend type, respectively; and $b^{(f)} + b^{(e)} = 1$. At each timeslot n, based on the current belief vector $\mathbf{b}_n$ and the auxiliary strategies $\boldsymbol{\pi}^{(e)}$ and $\boldsymbol{\pi}^{(f)}$ given by (4.25) and (4.29), user-I can derive the weighted strategy as follows

$$\boldsymbol{\pi}_n^I = b_n^{(e)} \cdot \boldsymbol{\pi}_n^{(e)} + b_n^{(f)} \cdot \boldsymbol{\pi}_n^{(f)}. \qquad (4.30)$$

In the meantime, user-I needs to build an estimated strategy for each possible type of user-II as suggested in the Bayesian Nash-Q algorithm. To this end, user-I can user $-Q_n^{(e)}$ and $Q_n^{(f)}$ as the estimates of the quality functions $Q_n^{II,(e)}$ and $Q_n^{II,(f)}$ for the enemy type and the friend type user-II, respectively. Based on these quality functions, user-I can further derive the most likely type-dependent strategies as follows

$$\hat{\pi}_{n+1}^{II,(e)}(s) = \arg\max_{\pi^{II}(s)} \min_{a^I \in \mathscr{A}^I} \sum_{a^{II}} Q_{n+1}^{II,(e)}(s, a^I, a^{II}) \pi_{a^{II}}^{II}(s), \qquad (4.31)$$

and

$$\hat{\pi}_{n+1}^{II,(f)}(s) = \arg\max_{\pi^{II}(s)} \max_{a^I \in \mathscr{A}^{\mathscr{I}}} \sum_{a^{II}} Q_{n+1}^{II,(f)}(s, a^I, a^{II}) \pi_{a^{II}}^{II}(s). \qquad (4.32)$$

Then, based on the Bayes' formula, the updated belief that user-II is a friend is given by

$$b_{n+1}^{(f)} = \frac{b_n^{(f)} \cdot f_n^{(f)}}{b_n^{(f)} \cdot f_n^{(f)} + b_n^{(e)} \cdot f_n^{(e)}}, \tag{4.33}$$

where the likelihoods $f_n^{(f)}$ and $f_n^{(e)}$ are given by

$$f_n^{(f)} = (1 - p_{explr}) \cdot \hat{\pi}_{a_n^{II},n}^{II,(f)}(s_n) + \frac{p_{explr}}{m}, \tag{4.34}$$

and

$$L_n^{(e)} = (1 - p_{explr}) \cdot \hat{\pi}_{a_n^{II},n}^{II,(e)}(s_n) + \frac{p_{explr}}{m}, \tag{4.35}$$

respectively. Accordingly, the belief for enemy type user-II is given by

$$b_{n+1}^{(e)} = 1 - b_{n+1}^{(f)}. \tag{4.36}$$

Unlike the dynamic IDS configuration problem presented in the previous subsection, the special structure property of the adversarial spectrum access problem allows one to establish the following convergence result.

Proposition 1 ([9]) *When user-I and user-II use the identical learning parameters, user-I's belief vector $\mathbf{b}_n$ converges to the true one. Additionally, the weighted strategy π_n^I given in (4.30) converges to the corresponding optimal strategy.*

Fig. 4.8 Type identification

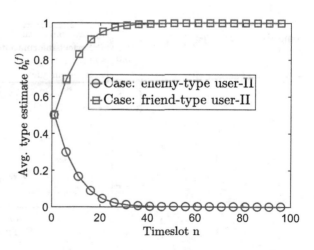

Now, let's examine the effectiveness of Bayesian Nash-Q algorithm in the adversarial spectrum access problem through some numerical explorations. Two cases where the true type of user-II is friend and enemy, respectively, are considered. The type identification results are shown in Fig. 4.8. It can be observed that, in both cases, the Bayesian Nash-Q algorithm enables user-I to achieve nearly perfect type identification after only about 40 timeslots. This can be explained by Proposition 1: Since the algorithm is guaranteed to converge, user-I can always make a perfect estimate about user-II's type-dependent strategies, and therefore, the Bayesian belief update procedure in Bayesian Nash-Q leads to the true belief. Similar to the dynamic IDS configuration problem, the performance of several different scenarios, "Bayesian method", "Perfect information", "Incorrect type", and "Random guess" are compared here. As it can be observed from Figs. 4.9 and 4.10 that, no matter what user-II's type is, the Bayesian Nash-Q algorithm always helps user achieve a nearly optimal performance after a sufficient number of timeslots.

Fig. 4.9 Throughput comparison: enemy-type user-II

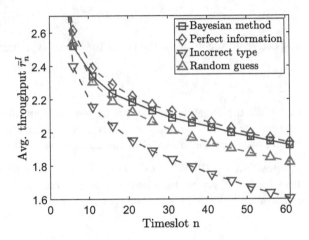

Fig. 4.10 Throughput comparison: friend-type user-II

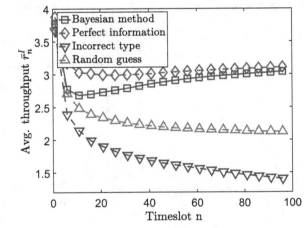

4.5 Summary

In this chapter, dynamic security games in which the defender lacks information about the attacker are investigated through the lens of Bayesian SG. To address such incomplete information SGs, the Bayesian Nash-Q algorithm, which is an integration of the repeated Bayesian games and the conventional Nash-Q algorithm, is studied. In addition, to illustrate the practical merit of the Bayesian SG framework and the Bayesian Nash-Q algorithm, their applications in addressing two important security issues, dynamic IDS configuration and spectrum access in adversarial environments, are examined. The corresponding numerical results validate the effectiveness of the Bayesian Nash-Q algorithm.

References

1. L. Busoniu, R. Babuska, B. De Schutter, A comprehensive survey of multiagent reinforcement learning. IEEE Trans. Syst. Man Cybern. C **38**(2), 156–172 (2008)
2. Q. Zhu, T. Basar, Dynamic policy-based IDS configuration, in *IEEE Conference on Decision and Control* (IEEE, New York, 2009), pp. 8600–8605
3. X. He, H. Dai, P. Ning, R. Dutta, Dynamic IDS configuration in the presence of intruder type uncertainty, in *2015 IEEE Global Communications Conference (GLOBECOM)* (IEEE, New York, 2015), pp. 1–6
4. A. Sampath, H. Dai, H. Zheng, B.Y. Zhao, Multi-channel jamming attacks using cognitive radios, in *Proceedings of 16th International Conference on Computer Communications and Networks, 2007 (ICCCN 2007)* (IEEE, 2007, New York), pp. 352–357
5. B. Wang, Y. Wu, K.J.R. Liu, T.C. Clancy, An anti-jamming stochastic game for cognitive radio networks. IEEE J. Sel. Areas Commun. **29**(4), 877–889 (2011)
6. Y. Wu, B. Wang, K.J.R. Liu, T.C. Clancy, Anti-jamming games in multi-channel cognitive radio networks. IEEE J. Sel. Areas Commun. **30**(1), 4–15 (2012)
7. H. Li, Z. Han, Dogfight in spectrum: combating primary user emulation attacks in cognitive radio systems, Part I: known channel statistics. IEEE Trans. Wirel. Commun. **9**(11), 3566–3577 (2010)
8. E. Altman, K. Avrachenkov, A. Garnaev, Jamming in wireless networks under uncertainty. Mob. Netw. Appl. **16**(2), 246–254 (2011)
9. X. He, H. Dai, P. Ning, R. Dutta, A stochastic multi-channel spectrum access game with incomplete information, in *Proceedings of IEEE ICC*, London, June (2015)
10. J.C. Harsanyi, Games with incomplete information played by "Bayesian players. Manag. Sci. **50**(12_supplement), 1804–1817 (2004)
11. R. Gibbons, *A Primer in Game Theory* (Harvester Wheatsheaf, New York 1992)
12. J.S. Jordan, Bayesian learning in repeated games. Games Econ. Behav. **9**(1), 8–20 (1995)
13. B. Mukherjee, L.T. Heberlein, K.N. Levitt, Network intrusion detection. IEEE Netw. **8**(3), 26–41 (1994)
14. I. Butun, S. Morgera, R. Sankar, A survey of intrusion detection systems in wireless sensor networks. IEEE Commun. Surv. Tutorials **16**(1), 266–282 (2014)
15. T.F. Lunt, A survey of intrusion detection techniques. Comput. Secur. **12**(4), 405–418 (1993)
16. H. Debar, M. Dacier, A. Wespi, Towards a taxonomy of intrusion-detection systems. Comput. Netw. **31**(8), 805–822 (1999)
17. S. Axelsson, Intrusion detection systems: a survey and taxonomy. Technical report, Department of Computer Engineering, Chalmers University, 2000

18. T. Anantvalee, J. Wu, A survey on intrusion detection in mobile ad hoc networks, in *Wireless Network Security* (Springer, Berlin, 2007), pp. 159–180
19. P. Garcia-Teodoro, J. Diaz-Verdejo, G. Maciá-Fernández, E. Vázquez, Anomaly-based network intrusion detection: techniques, systems and challenges. Comput. Secur. **28**(1), 18–28 (2009)
20. N.S. Evans, R. Dingledine, C. Grothoff, A practical congestion attack on Tor using long paths, in *USENIX Security Symposium*, pp. 33–50 (2009)
21. M.L. Littman, Markov games as a framework for multi-agent reinforcement learning, in *Proceedings of ICML*, New Brunswick, NJ (1994)

Chapter 5
Dynamic Security Games with Deception

5.1 Introduction

In the previous two chapters, we have investigated dynamics security games with asymmetric information, in which the extra or the missing information is inherent to the underlying security problems.

In this chapter, we switch the gear and further consider the possibility of proactively creating information asymmetry in security games for the defender's benefit. Particularly, deception will be investigated as a concrete tool to achieve this objective. Actually, deception has already been widely considered in literature to address different security related issues. For example, in [1–4], deception mechanisms have been examined in the context of military and anti-terrorism applications. In addition, cyber space deception techniques have also been employed to protect various information systems and networks [4–9]. The essential idea of deception is that the defender can spend some security resource to proactively create certain faked information so as to mislead the adversary and achieve a better defense. However, most existing deception techniques are designed for static scenarios. The objective of this chapter is to illustrate a foresighted deception framework that allows the defender to better align its deception strategy with its conventional defense strategy and the dynamics in the environment.

The desired foresighted deception technique is fulfilled through incorporating an additional deception module into the conventional SG model, resulting in a new stochastic deception game (SDG) framework. To enable the defender to find good deception and defense strategies, a computing algorithm and a learning algorithm are presented, respectively. The computation algorithm is mainly designed for the offline settings where the defender has full information about the SDG while the learning algorithm is designed for the online settings where the defender lacks information about the system and environmental dynamics. In addition, a network

© The Author(s) 2018
X. He, H. Dai, *Dynamic Games for Network Security*, SpringerBriefs in Electrical and Computer Engineering, https://doi.org/10.1007/978-3-319-75871-8_5

protection game is taken as a concrete example to illustrate the advantage of the foresighted deception achieved by the SDG framework over the conventional myopic deception.

5.2 Stochastic Deception Games

In this section, the stochastic deception game (SDG) model that allows the defender to conduct active deception during the course of dynamic security competitions will be introduced first, followed by two novel algorithms that enable the defender to compute and learn the best possible defense and deception strategies.

5.2.1 The SDG Model

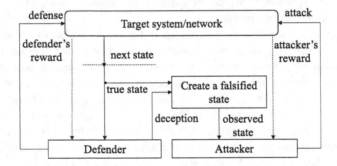

Fig. 5.1 Diagram of SDG

The SDG is developed on top of the conventional SG introduced in Chap. 1 and further incorporates a deception procedure. As illustrated in Fig. 5.1, in addition to those steps taken in conventional SG, the defender in SDG further takes a deception action that actively creates a falsified state $g(s_n) \in \mathscr{S} \cup \{\emptyset\}$ to mislead the attacker so as to achieve a better protection of the target system. Particularly, let $d_n(s_n)$ denote the amount of security resource devoted to fulfill the intended deception, and the corresponding cost of deception is denoted by $C^D(d_n)$; for illustration purpose, a simple cost function $C^D(d_n) = d_n$ is assumed in the discussions hereafter. Apparently, the more security resource is devoted, the higher success rate of the deception is. To facilitate the analysis, it is assumed that, when $d_n(s_n)$ units of security resource is devoted to deception, the attacker will observe the falsified state with a probability given by

$$\mathbb{P}(d_n) = 1 - \exp(-\lambda d_n), \tag{5.1}$$

where the parameter $\lambda > 0$ is the efficiency coefficient of the deception technique adopted by the defender; to achieve the same $\mathbb{P}(d_n)$, less security resource is required when λ is larger. It is worth mentioning that the above probability model is used only for illustration purpose and the SDG works for other forms of probability models as well. Accordingly, with probability $1 - \mathbb{P}(d_n)$, the deception fails and the attacker observes the true state s_n. In the following discussion, it is assumed that the defender always creates a null state $g(s_n) = \emptyset$ during deception to prevent the attacker from knowing the true state of the target system, and more advanced falsified state selection strategy is beyond the scope of the discussion here. In the SDG, the reward of the defender depends on both the security reward, which is given by

$$r_n^D = R^D(s_n, a_n^D, a_n^A), \tag{5.2}$$

and the deception cost $C^D(d_n)$. Consequently, in the SDG, the long term performance to be optimized by the defender is given by $\mathbb{E}\left[\sum_{n=1}^{\infty} \beta^{n-1} \left(r_n^D - C^D(d_n)\right)\right]$. By further taking the zero-sum assumption, it is clear that the attacker's instant reward at each timeslot is given by $r_n^A = -r_n^D + C^D(d_n)$. Note that the SDG may be treated as a generalization of the conventional myopic deception (corresponding to the case of $\beta = 0$). As to the attacker, due to deception, it can only launch attacks based on its observed state $\hat{s}_n$ and when deception succeeds and the observed state is the null state $\hat{s} = \emptyset$, it is assumed that the attacker will take an action a^A selected uniformly at random.

In the SDG described above, the defender has to find its optimal defense strategy as in the conventional SG and also the optimal deception d_n. To this end, some new notations have to be introduced to facilitate later analysis. First, let us denote by $\tilde{V}^D(s, g(s))$ and $\tilde{V}^D(s, s)$ the defender's long-term performance when the deception succeeds and fails, respectively, excluding the deception cost at the current time. These two quantities will be termed as the *intermediate value functions*, so as to differentiate from the conventional value function $V^D(s)$. By their definitions, it is not difficult to realize the following relation among these value functions:

$$V^D(s) = -d(s) + \exp(-\lambda \cdot d(s)) \cdot \tilde{V}^D(s, s) + (1 - \exp(-\lambda \cdot d(s))) \cdot \tilde{V}^D(s, g(s)). \tag{5.3}$$

Let us further define the Q-function $\tilde{Q}^D(s, a^D, a^A)$ for the SDG according to the well-known Bellman equation as follows:

$$\tilde{Q}^D(s, a^D, a^A) = R^D(s, a^D, a^A) + \beta \cdot \sum_{s' \in \mathscr{S}} \mathbb{P}(s'|s, a^D, a^A) \cdot V^D(s'), \quad (5.4)$$

which holds similar physical interpretations as the Q-function in conventional SG. It is not difficult to realize that, when deception fails, the SDG reduces to the conventional SG, and hence both the defender and the attacker will follow their own state-dependent strategies at NE. Consequently, the corresponding intermediate value function admits

$$\tilde{V}^D(s, s) = \mathrm{NE}\left(\tilde{Q}^D, s, s\right). \quad (5.5)$$

In the above equation, the operator NE() outputs the value of a zero-sum game with a payoff matrix specified by $\tilde{Q}^D$ and s. When deception succeeds, the defender can do better than simply follows the NE. Instead, since the attacker cannot observe the true state and take a random action, the defender can take the best response against such random strategy of the attacker. It is not difficult to realize that the corresponding value to the defender is given by

$$\tilde{V}^D(s, g(s)) = \mathrm{BR}\left(\tilde{Q}^D, s, g(s)\right), \quad (5.6)$$

where BR() is the operator of computing the value of a game at best response [10]. Summarizing the above discussion, the defender has to solve the following optimization problem:

$$\max_{\{d(s^i)\}_{i=1}^{|\mathscr{S}|} \geq 0} \left(V^D(s^1), \ldots, V^D(s^{|\mathscr{S}|})\right) \qquad \textbf{(P)}$$

$$\text{s.t.} \quad (C1) \; V^D(s) = -d(s) + \exp(-\lambda d(s)) \cdot \tilde{V}^D(s, s)$$

$$+ (1 - \exp(-\lambda d(s))) \cdot \tilde{V}^D(s, g(s)), \; \forall s \in \mathscr{S}$$

$$(C2) \; \tilde{Q}^D(s, a^D, a^A) = R^D(s, a^D, a^A)$$

$$+ \beta \cdot \sum_{s' \in \mathscr{S}} \mathbb{P}(s'|s, a^D, a^A) \cdot V^D(s'), \; \forall s \in \mathscr{S}, \forall a^D \in \mathscr{D}, \forall a^A \in \mathscr{A},$$

$$(C3) \; \tilde{V}^D(s, s) = \mathrm{NE}\left(\tilde{Q}^D, s, s\right), \; \forall s \in \mathscr{S},$$

$$(C4) \; \tilde{V}^D(s, g(s)) = \mathrm{BR}\left(\tilde{Q}^D, s, g(s)\right), \; \forall s \in \mathscr{S}.$$

However, as the above optimization problem is non-linear and non-convex and involves with multi-objective and equilibrium constraints. To the best of our knowledge, finding the global optimal solution for such an optimization problem still remains an open problem. With this consideration, studying algorithms that can find a reasonably good solution to this optimization problem will be the focus of our discussion below.

5.2.2 Solving the SDG

In this section, an iterative computing algorithm (Algorithm 1) is presented to solve the SDG when the state transition probability is known to the defender first. In addition, to cope with unknown environmental dynamics, a corresponding learning algorithm (Algorithm 2) is presented as well. With these two algorithms, the defender can effectively conduct desirable foresighted deception for enhanced security performance.

As demonstrated in Fig. 5.2, the computing algorithm iteratively evaluates the values of the deception strategy d, V^D, $\tilde{Q}^D$, and $\tilde{V}^D$. More specifically, based on the intermediate value functions $V^D_{n-1}(s, s)$ and $\tilde{V}^D_{n-1}(s, g(s))$ obtained in the last iteration, the optimal deception can be found by differentiating (5.3) with respective to d, which gives

$$d_n(s) = \max \left\{ 0, \frac{1}{\lambda} \log \left(\lambda \cdot \left(\tilde{V}^D_{n-1}(s, g(s)) - \tilde{V}^D_{n-1}(s, s) \right) \right) \right\}. \qquad (5.7)$$

Based on the updated deception strategy d_n, the value function of the SDG is then updated as follows

Fig. 5.2 Diagram of Algorithm 1

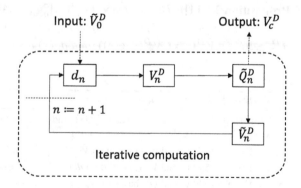

Input: $\tilde{V}^D_0$ Output: V^D_c

$d_n \longrightarrow V^D_n \longrightarrow \tilde{Q}^D_n$

$n := n+1$

$\tilde{V}^D_n$

Iterative computation

$$V_n^D(s) = -d_n(s) + e^{-\lambda d_n(s)} \cdot \tilde{V}_{n-1}^D(s,s) + \left(1 - e^{-\lambda d_n(s)}\right) \cdot \tilde{V}_{n-1}^D(s, g(s)), \quad (5.8)$$

The next step is to update the Q-function based on the updated value function V_n^D as follows

$$\tilde{Q}_n^D(s, a^D, a^A) = R^D(s, a^D, a^A) + \beta \sum_{s' \in \mathscr{S}} \mathbb{P}(s'|s, a^D, a^A) \cdot V_n^D(s'). \quad (5.9)$$

Finally, the intermediate value functions are updated based on the updated Q-function as follows

$$\tilde{V}_n^D(s,s) = \text{NE}(\tilde{Q}_n^D, s, s), \quad (5.10)$$

$$\tilde{V}_n^D(s, g(s)) = \text{BR}(\tilde{Q}_n^D, s, g(s)). \quad (5.11)$$

At each iteration, the corresponding optimal defense strategy of the defender can be computed based on $\tilde{Q}_n^D$. Particularly, when deception fails, the defender takes the strategy

$$\pi_{\text{NE},n}^D(s, \cdot) = \arg \text{NE}(\tilde{Q}_n^D, s, s), \quad (5.12)$$

and when the deception succeeds, it follows

$$\pi_{\text{BR},n}^D(s, \cdot) = \arg \text{BR}(\tilde{Q}_n^D, s, g(s)), \quad (5.13)$$

The above steps are summarized in Algorithm 1, and the its convergence property is established by the following proposition.

Proposition 2 ([11]) *The sequence $\{V_n^D(s)\}_{n=0}^{\infty}$ in Algorithm 1 converges monotonically to a point $V_c^D(s)$. Consequently, other relevant quantities $\tilde{Q}_n^D(s, \cdot, \cdot)$, $\pi_{\text{NE},n}^D$, $\pi_{\text{BR},n}^D$, $\tilde{V}_n^D(s,s)$, $\tilde{V}_n^D(s, g(s))$, and $d_n(s)$ also converge.*

Algorithm 1 The computing algorithm

1: Initialize $\tilde{V}_0^D(s,s)$, $\tilde{V}_0^D(s, g(s))$ and $V_n^D(s)$, and start with $n = 1$.
2: **while** $n > 1$ and $|V_n^D(s) - V_{n-1}^D(s)| > \epsilon$ **do**
3:    Execute (5.7) to find $d_n(s)$ based on $\tilde{V}_{n-1}^D(s, g(s))$ and $\tilde{V}_{n-1}^D(s,s)$;
4:    Execute (5.8) to find $V_n^D(s)$ based on $d_n(s)$, $\tilde{V}_{n-1}^D(s, g(s))$ and $\tilde{V}_{n-1}^D(s,s)$;
5:    Execute (5.9) to find $\tilde{Q}_n^D$ based on $V_n^D(s)$;
6:    Execute (5.10) and (5.11) to find $\tilde{V}_n^D(s,s)$ and $\tilde{V}_n^D(s, g(s))$ based on $\tilde{Q}_n^D$;
7:    Execute (5.12) and (5.13) to find $\pi_{\text{NE},n}^D$ and $\pi_{\text{BR},n}^D$;
8:    Let $n := n + 1$;
9: **end while**.

The computation algorithm introduced above is only suitable to the scenarios where the state transition probability $\mathbb{P}(s'|s, a^D, a^A)$ is known to the defender. To cope with security problems with unknown dynamics, a learning algorithm that enables the defender to learn its defense and deception strategies is presented below. This algorithm is built by incorporating a reinforcement learning procedure similar to [12] into Algorithm 1. Particularly, the computation dictated by (5.9) will be replaced by the following learning process

$$
\tilde{Q}_n^D(s, a^D, a^A) =
\begin{cases}
(1 - \alpha_n)\tilde{Q}_{n-1}^D(s, a^D, a^A) + \alpha_n \left(R^D(s, a^D, a^A) + \beta V_{n-1}^D(s_{n+1}) \right) \\
\qquad\qquad\qquad\qquad\qquad , \text{ for } (s, a^D, a^A) = (s_n, a_n^D, a_n^A) \\
\tilde{Q}_{n-1}^D(s, a^D, a^A), \qquad\qquad\qquad\qquad\qquad \text{otherwise.}
\end{cases}
$$

The overall learning algorithm is summarized in Algorithm 2, and its convergence property is given by Proposition 3.

Proposition 3 ([11]) *All the quantities in Algorithm 2 converge to their corresponding converging points given by Algorithm 1.*

5.3 A Security Application

A network protection game depicted in Fig. 5.3 will be taken as an example to illustrate the application of the SDG presented in the previous section. Specifically, in the network protection game, there is a target network consisting of multiple interconnected computers. The objective of the defender is to keep as many computers healthy as possible, whereas the attacker aims to bring down as much computers as possible by injecting malware into the computers. In this problem, it is

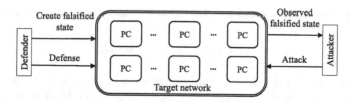

Fig. 5.3 The network protection game

assumed that the injected malware may spread from infected computers to healthy computers with a certain (possibly unknown) probability, causing state dynamics in the target network. In addition to regular defense (e.g., regular patching, system updating, etc.), the defender can spend some security resource to prevent the attacker from knowing which computers are infected and which are not. If such deception

---

Algorithm 2 The learning algorithm

1: Initialize $\tilde{Q}(s, \cdot, \cdot)$, $\tilde{V}_0^D(s, s)$, $\tilde{V}_0^D(s, g(s))$, $V_n^D(s)$, $\pi_{\text{NE},0}^D(s, \cdot)$ and $\pi_{\text{BR},0}^D(s, \cdot)$, and start with
   $n = 1$.
2: **while** not converge **do**
3:     Observe s_n;
4:     Execute (5.7) and (5.8) to find $d_n(s_n)$ and $V_n^D(s_n)$;
5:     Allocate $d_n(s_n)$ units of resource for deception;
6:     Generate $x_n \in [0, 1]$ uniformly at random;
7:     **if** deception succeeds **then**
8:         **if** $x_n \leq p_{explr}$ **then**
9:             Take a defense action a_n^D chosen uniformly at random;
10:         **else**
11:             Take a defense action a_n^D with probability $\pi_{\text{BR},n-1}^D(s_n, a_n^D)$;
12:         **end if**
13:     **else**
14:         **if** $x_n \leq p_{explr}$ **then**
15:             Take a defense action a_n^D chosen uniformly at random;
16:         **else**
17:             Take a defense action a_n^D with probability $\pi_{\text{NE},n-1}^D(s_n, a_n^D)$;
18:         **end if**
19:     **end if**
20:     Execute (5.14) to find $\tilde{Q}_n$;
21:     Execute (5.10) and (5.11) to find $\tilde{V}_n^D(s, s)$ and $\tilde{V}_n^D(s, g(s))$;
22:     Execute (5.12) and (5.13) to find $\pi_{\text{NE},n}^D$ and $\pi_{\text{BR},n}^D$;
23:     Let $n := n + 1$;
24: **end while**

---

succeeds, the attacker may falsely inject malware on an already infected computer,
reducing its attacking efficiency.

To apply the SDG to address the above security problem, one may model the
defender's reward function as follows

$$r_n^D = R^D(s_n, a_n^D, a_n^A) = \varphi(k_n) - \varphi_D \cdot |a_n^D| + \varphi_A \cdot |a_n^A|, \qquad (5.14)$$

where

$$\varphi(k_n) = \varphi_0 \cdot \mathbb{1}_{\{k_n > 0\}} + \Delta\varphi \cdot k_n. \qquad (5.15)$$

Some explanations of the above reward function are in order. The first term $\varphi(k_n)$
denotes the reward to the defender when k_n computers are healthy in the target
number. The second and the third terms correspond to the defender's defense cost
and the attacker's attack cost, respectively. More specifically, a_n^D and a_n^A are the
actions taken by the defender and the attacker, respectively, representing the set of
computers they choose to act on. Further taking a zero-sum assumption, we have
$r_n^A = -r_n^D$.

The performance of the SDG on this problem is illustrated in Figs. 5.4, 5.5, and
5.6. Particularly, the convergence of the value functions in Algorithm 1 is illustrated

in Fig. 5.4. It can be seen that all the value functions increase monotonically and eventually converge to a stable value, as predicted by Proposition 2. The convergence of Algorithm 2 is validated in Fig. 5.5. It can be seen that, as predicated by Proposition 3, the value functions learned through Algorithm 2 eventually converge to the same value as in Algorithm 1. The average security reward, defined as

$$\bar{r}_n^{D,\beta} \triangleq \frac{1}{n} \sum_{i=1}^{n} r_i^D - C^D(d_i),\tag{5.16}$$

will be used as the metric to measure the performance in this problem. In Fig. 5.6, we will compare the security performance of Algorithm 1 and that of the conventional myopic deception (corresponding to $\beta = 0$ in the SDG). To this end, the relative performance gain over the myopic deception, defined as

$$\eta \triangleq \frac{\bar{r}_T^{D,\beta} - \bar{r}_T^{D,0}}{\bar{r}_T^{D,0}} \times 100\%,\tag{5.17}$$

is shown in Fig. 5.6 for different β's and deception coefficient λ's. It can be seen that, by following the strategy derived from the SDG, the defender can achieve a substantially better performance as compared to the myopic deception case. For example, when the discounting factor is set to $\beta = 0.9$ and the deception coefficient is $\lambda = 1.5$, the strategy derived from SDG is more than four times better than that of the conventional myopic deception. Also, larger discounting factor β and deception coefficient λ correspond to better performance. This is because, when the discounting factor is larger, the defender conducts a more foresighted optimization, and when the deception coefficient is larger, less security resource is required to achieve the same deception success probability.

5.4 Summary

In this chapter, we have studied how the defender can proactively create faked system states to mislead the adversary through deception. To achieve foresighted deception in dynamic environments, the SDG framework is introduced in details and the corresponding computing and learning algorithms are also presented to enable the defender to find a good deception strategy in offline and online settings, respectively. In addition, through an exemplary application of the network protection game, the appealing performance of the SDG framework is illustrated.

Fig. 5.4 Convergence of
value functions in
Algorithm 1

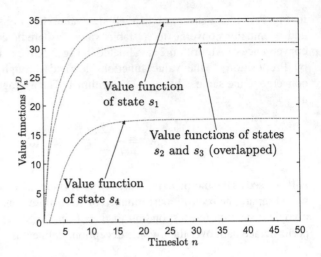

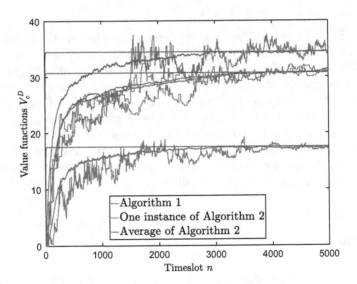

Fig. 5.5 Convergence of value function in Algorithm 2

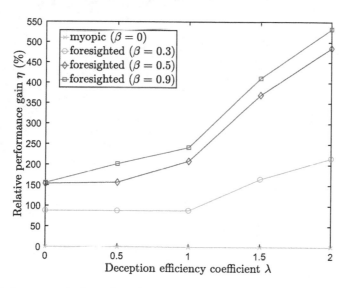

Fig. 5.6 Performance gain due to deception

References

1. J.P. Hespanha, Y. Ateskan, H. Kizilocak, Deception in non-cooperative games with partial information, in *DARPA-JFACC Symposium on Advances in Enterprise Control* (2000)
2. D. Li, J.B. Cruz, Information, decision-making and deception in games. Decis. Support. Syst. **47**(4), 518–527 (2009)
3. J. Zhuang, V.M. Bier, Secrecy and deception at equilibrium, with applications to anti-terrorism resource allocation. Def. Peace Econ. **22**(1), 43–61 (2011)
4. J. Zhuang, V.M. Bier, O. Alagoz, Modeling secrecy and deception in a multiple-period attacker–defender signaling game. Eur. J. Oper. Res. **203**(2), 409–418 (2010)
5. T.E. Carroll, D. Grosu, A game theoretic investigation of deception in network security. Secur. Commun. Netw. **4**(10), 1162–1172 (2011)
6. J. Pawlick, Q. Zhu, Deception by design: evidence-based signaling games for network defense (2015). arXiv preprint arXiv:1503.05458
7. H. Xu, Z. Rabinovich, S. Dughmi, M. Tambe, Exploring information asymmetry in two-stage security games, in *Proceedings of AAAI* (2015)
8. F. Cohen, D. Koike, Misleading attackers with deception, in *Proceedings of IEEE SMC Information Assurance Workshop* (2004)
9. H. Xu, A.X. Jiang, A. Sinha, Z. Rabinovich, S. Dughmi, M. Tambe, Security games with information leakage: modeling and computation (2015). arXiv preprint arXiv:1504.06058
10. T. Alpcan, T. Başar, *Network Security: A Decision and Game-Theoretic Approach* (Cambridge University Press, Cambridge, 2010)
11. X. He, M.M. Islam, R. Jin, H. Dai, Foresighted deception in dynamic security games, in *2017 IEEE International Conference on Communications (ICC)*, pp. 1–6
12. M.L. Littman, Markov games as a framework for multi-agent reinforcement learning, in *Proceedings of ICML* (1994)

Chapter 6
Conclusion and Future Work

6.1 Summary

In this book, we have studied the applications of SG theory in addressing various dynamic network security games, with a particular focus on the techniques of handling scenarios with information asymmetry.

In Chap. 1, to pave the way for later discussions, some preliminaries of game theory, MDP, and SG have been reviewed, including the basic concepts, the mathematical models, and the corresponding solution techniques. With these necessary backgrounds, a brief review of existing applications of the SG framework in analyzing the strategic interactions between the defenders and the attackers in different network security competitions is provided in Chap. 2. Particularly, cyber networks, wireless communication networks, and cyber-physical networks are the focuses of the discussions there. However, in these existing applications, equal information is often assumed available to the defenders and the attackers, whereas in practical network security problems, information asymmetry exists. With this consideration, we have further examined several novel techniques for SG with information asymmetry and their security applications. Particularly, the discussions in Chap. 3 focus on dynamic security games with extra information. To enable the defender to fully exploit its information advantage, the PDS-learning technique has been incorporated into conventional multi-agent RL algorithms, leading to two new algorithms, minimax-PDS and WoLF-PDS. The EHCS anti-jamming problem and the cloud-based security game are taken as two examples to illustrate the effectiveness of these two algorithms. The complementary situations where the defender only has incomplete information about the ongoing security rivalries are examined in Chap. 4 through the lens of Bayesian SG. To this end, the Bayesian Nash-Q algorithm, an integration of the repeated Bayesian games and the conventional Nash-Q algorithm, has been studied in details. To illustrate the practical merit of this algorithm, its applications in dynamic IDS configuration and spectrum access in adversarial environments have been presented. The extra

© The Author(s) 2018
X. He, H. Dai, *Dynamic Games for Network Security*, SpringerBriefs in Electrical and Computer Engineering, https://doi.org/10.1007/978-3-319-75871-8_6

and the incomplete information considered in Chaps. 3 and 4 is inherent to the corresponding security problems. In Chap. 5, we have studied how the defender can proactively create information asymmetry to enhance security performance. For this purpose, the SDG framework has been studied along with the corresponding computing and learning algorithms. The performance of the SDG framework was examined through an exemplary network protection game.

6.2   Future Works

Based on the discussions in this book, it can be realized that the SG theory provides a solid theoretic foundation for analyzing various existing and emerging dynamic network security games. Nonetheless, to better capture the nuance in practical security problems for enhanced security performance, further advancement of the conventional SG theory is needed.

From the theoretic viewpoint, the following research problems remain interesting. Firstly, a *quantitative analysis* of the security gain that can be brought by the extra information through the algorithms discussed in Chap. 3 is still missing. The merit of such quantitative analysis is twofold. On the one hand, it enables a more precise prediction about the performance in dynamic security games with extra information. On the other hand, it can provide a measure of the importance of information so as to guide the defender towards better information protection. In addition, it is also interesting to consider the possibility of *incorporating the PDS-learning principle into other multi-agent RL algorithms* (e.g., Nash-Q and FoF-Q) and explore their security applications. As to the incomplete information SG discussed in Chap. 4, it is worth further investigating the situations with more intelligent attackers who may deliberately switch its types over time so as to prevent the defender from identifying its real purpose. Moreover, to make the SDG framework presented in Chap. 5 more complete, one has to further consider the situations where the attacker may also spend some extra resource to conduct surveillance against the defender's deception. Lastly, pursuing an integrated framework for addressing all the above-mentioned issues will be a challenging but rewarding task.

From the application viewpoint, the first desirable objective is to *scale up existing algorithms* to handle practical security problems with much larger state and action spaces as well as multiple defenders and attackers. Reducing computational cost and shortening learning time in those algorithms are also important issues to be addressed; one possible solution is to exploit the recent advancement is parallel computing. In addition, it is crucial to *examine the performance of these theoretic outcomes in real-world networks and security systems*. For example, one may consider the implementing the Bayesian Nash-Q algorithm to configure real-world IDS and examining its performance. Instantiation of the deception techniques in real-world cyber-physical systems also worth further investigations.

Printed in the United States
By Bookmasters